THERMODYNAMIC AND TRANSPORT PROPERTIES OF COAL LIQUIDS

EXXON MONOGRAPHS

THERMODYNAMIC AND TRANSPORT PROPERTIES OF COAL LIQUIDS

CONSTANTINE TSONOPOULOS
JOHN L. HEIDMAN
SHUEN-CHENG HWANG

An Exxon Monograph

JOHN WILEY & SONS
New York • Chichester • Brisbane • Toronto • Singapore

Copyright © 1986 by John Wiley & Sons, Inc.

All rights reserved. Published simultaneously in Canada.

Library of Congress Cataloging in Publication Data:

Tsonopoulos, Constantine.
    Thermodynamic and transport properties of coal
liquids.

    "An Exxon monograph."
    Includes bibliographical references and index.
    1. Coal liquefaction. 2. Thermodynamics. 3. Mass
transfer. I. Heidman, John L. II. Hwang, Shuen-Cheng.
III. Title.
TP352.T78 1986   662'.669   85-17912
ISBN 0-471-83282-0

Printed in the United States of America

10 9 8 7 6 5 4 3 2 1

PREFACE

Coal liquefaction has attracted attention in the last few years because it is expected to provide a significant part of the liquid fuels of the future. Several processes are being investigated, and in each of them process development has been hampered by the unavailability of thermodynamic and transport property data at the conditions of interest. Most of these properties have to be predicted with methods based on low-temperature data for the predominantly paraffinic and naphthenic petroleum liquids. It is our goal in this book to demonstrate how well we can extend and modify these methods to make them applicable to the highly aromatic coal liquids, that is, to the high temperatures needed to liquefy coal.

This book is limited in scope. We have not attempted to make an exhaustive examination of what is known about all coals and all coal liquids. Practically nothing is said about the processes, except for the conditions of interest. We have focused on the thermodynamic and transport properties needed in heat and material balances, as well as equipment design calculations. Recent research, at Exxon and elsewhere, involving experimental work, theoretical analysis, and correlation development has considerably expanded our knowledge of the properties of coal liquids. We confess that we have emphasized what we know best: the EDS process and the three coals that were used in the large-scale demonstration of this process (Illinois No. 6, Wyoming Wyodak, and, to a lesser extent, Texas lignite). But we have also included the extensive data on SRC-II liquids that were published by J. A. Gray of Gulf Research and Development Company. SRC-I, H-Coal, and other direct coal liquefaction processes are represented to a more limited extent because much less has been published about them.

The predictive methods we emphasize in this book are those used in the industry for petroleum liquids. Two important publications summarize most of these methods: J. B. Maxwell, *Data Book on Hydrocarbons*, Van Nostrand, Princeton, NJ, 1950; American Petroleum Institute, *Technical Data Book—Petroleum Refining*, 4th Ed., API, Washington, DC, 1983. Maxwell's book is the 1950 version of the Exxon Data Book. Although we favor the methods used at Exxon, we also include comparisons with other approaches. For many properties, we use two broad methods: one based on boiling point and specific gravity, and the other on critical constants and the acentric factor.

The first chapter describes the key differences between coal liquids and petroleum liquids. It also introduces aromatic compounds that we selected as model compounds for coal liquids. Data needs for process design are reviewed in Chapter 2—first, generally, then for all coal liquefaction processes, and, finally, for the EDS process. Data and methods for characterizing coal liquids and model compounds are presented in Chapter 3. Chapters 4 through 9 deal with the properties of interest, especially at high temperatures: vapor pressure (4), vapor-liquid equilibria (5), thermal properties (6), density (7), surface tension (8), and transport properties (9). The book concludes with a look in Chapter 10 at future needs, followed by two brief appendices.

We have emphasized the properties of liquids. Very little is included about properties of gases (which, however, can generally be predicted reliably with existing correlations). There is nothing on the properties of coal—readers can refer to the appropriate sections in the DOE/IGT *Coal Conversion Systems Technical Data Book*—and only a brief discussion of coal/liquid slurries. A separate monograph is probably warranted on the properties of such slurries, a key concern in the direct coal liquefaction processes.

We envision this as a reference book for the student, the specialist, and, especially, the engineer working on coal liquefaction process development and design. Along with the publications mentioned earlier, it can be considered a companion to the standard reference for anyone involved in property predictions—R. C. Reid, J. M. Prausnitz, and T. K. Sherwood, *The Properties of Gases and Liquids*, 3rd Ed., McGraw-Hill, New York, 1977.

Al Schriesheim suggested that we write this book. Ben Weil, first editor of Exxon Monographs, helped us in our initial steps. We also wish to acknowledge the help of the assistant editor, Joyce Barrett, who, along with John Hack, reviewed the manuscript. Fred Horowitz, Stan Whittingham, Ted Kalina, Larry Kaye, Barry Tarmy, and, especially, Joe O'Bara helped us along and gave us the support we needed. We had very helpful discussions with our colleagues Paul Becker, Ralph Gray, Gerry Lahn, Pat Madden, Jerry Melin, and Gerry Vick. The manuscript was typed by Susan Flood.

The EDS coal liquid samples were obtained and analyzed at the Baytown laboratories of Exxon Research and Engineering Company; some analyses were also run at the Linden laboratories. We should mention George Anderson and John Taunton, among many others, for their help. Data on vapor pressure, vapor-liquid equilibria, density, surface tension, and viscosity were obtained at the Thermochemical Institute of Brigham Young University and Wiltec Research Company of Provo, Utah. Grant Wilson (at BYU and Wiltec) and John Cunningham (at BYU) were in charge of the experimental work. Heat capacity measurements were made at the Linden laboratories of ER&E by Deborah O'Rourke and Steve Mraw, while heats of combustion were measured at Baytown by Frank Tao.

Data on the properties of model compounds up to the high temperatures of interest have increased substantially in recent years, thanks to the work of K. C. Chao, A. J. Kidnay, R. Kobayashi, the National Institute for Petroleum and Energy Research (the new name for the Bartlesville Energy Technology Center), and others. The National Bureau of Standards is developing a high-temperature experimental capability. We hope that our book will encourage more experimental and theoretical work. More compelling reasons for doing such work are given in the U.S. Department of Energy report on "Thermophysical Properties for Synthetic Fuels," J. Kestin, Ed., November 1982.

CONSTANTINE TSONOPOULOS
JOHN L. HEIDMAN
SHUEN-CHENG HWANG

Florham Park, New Jersey
January 1986

ABOUT THE AUTHORS

Costa Tsonopoulos is a Senior Engineering Associate in the Technology Department of Exxon Research and Engineering Company in Florham Park, New Jersey. He received a doctoral degree in chemical engineering from the University of California at Berkeley in 1970. In March of that year, he joined Exxon and has spent all of his career in the Applied Thermodynamics Group, becoming group head in 1979. Dr. Tsonopoulos began his research on the properties of coal liquids in 1974. He has published 20 papers and has one patent. Among other outside activities, he is Exxon's representative on the American Petroleum Institute's Technical Data Committee and the American Institute of Chemical Engineers' Design Institute for Physical Property Data. In 1981–1982, Dr. Tsonopoulos was a member of a Department of Energy working group on thermophysical properties of synthetic fuels.

John L. Heidman is a Staff Engineer in the Technology Department of Exxon Research and Engineering Company in Florham Park, New Jersey. He earned both bachelor's (1978) and master's (1979) degrees in chemical engineering at Oklahoma State University. Since 1979, he has worked in the Applied Thermodynamics Group, where his major research efforts have focused on modeling phase equilibrium behavior and the thermal properties of petroleum and synthetic liquids. He has coauthored several papers on these subjects, and is a member of the American Institute of Chemical Engineers and the American Chemical Society.

Shuen-Cheng Hwang is a Senior Engineer at BOC Group, Inc., where he is responsible for technical data compilation and correlation for gas process engineering applications. Before his association with BOC, he worked in the Applied Thermodynamics Group of Exxon Research and Engineering Company, where he conducted research on the thermophysical properties of coal liquids. Dr. Hwang studied chemical engineering at National Cheng Kung University (B.S., 1967) and Oklahoma State University (M.S., 1972; Ph.D., 1975). He is a member of the American Institute of Chemical Engineers, the American Chemical Society, and the National Society of Professional Engineers.

CONTENTS

SYMBOLS

A_i = molar area of component i (Eq. 10.9)

$\bar{A}_i$ = partial molar area of component i (Eq. 10.9)

a, b = parameters in the Redlich-Kwong equation of state

$^\circ API$ = API gravity $= 141.5/S - 131.5$

$c_{i(j)}$ = characteristic binary constant in density-dependent mixing rule for a_m (Eq. 10.8)

C_{ij} = characteristic binary constant in Redlich-Kwong equation of state (Eq. 5.12)

C_p = heat capacity at constant pressure per unit mass (Btu/lb · °F)

D_{ij} = characteristic binary constant in Redlich-Kwong equation of state (Eq. 5.15)

$D_{1,2}$ = diffusion coefficient for 1 into 2

f_i = fugacity of component i

G = Gibbs free energy

H = enthalpy

k_T = thermal conductivity [Btu/(h · ft · °F)]

K_i = K-value of component $i = y_i/x_i$

K_w = Watson characterization factor $= [T_b(^\circ R)]^{1/3}/S$

M = molecular weight

P = pressure (psia; sometimes in atm or mm Hg)

P^* = parachor (Eq. 8.2)

R = gas constant

$\bar{R}$ = radius of gyration (Chapter 9; Eq. 9.13)

S = specific gravity at 60/60°F

t = temperature (°F; sometimes in °C)

T = absolute temperature (°R; sometimes in K)

v = molar volume

V = total volume

v^* = characteristic volume in COSTALD liquid density correlation (Eq. 7.6a)

w_i = weight fraction of component i

x_i = liquid mole fraction of component i

y_i = vapor mole fraction of component i

z_i = mole fraction of component i (phase unspecified)

Z = compressibility factor $= Pv/RT$

Z_{RA} = constant in Rackett liquid density correlation (Eq. 7.3)

GREEK SYMBOLS

α = polarizability (Eq. 10.5)

β = isothermal compressibility $= -(1/V)(\partial V/\partial P)_T$

γ_i = activity coefficient of component i

ΔX_a = change in property X upon transformation a

μ = dynamic viscosity (Ch. 9; in cP)

μ = dipole moment (Eq. 10.5)

μ' = effective dipole moment (Eq. 10.5)

ν = kinematic viscosity $= \mu/\rho$ (in cSt)

ρ = density (molar; sometimes mass)

σ = surface tension (in dyn/cm)

ϕ_i = fugacity coefficient of component $i = f_i/z_i P$

Φ_i = volumetric fraction of component i (Eq. 7.13; slightly different definition in Eq. 9.11c)

ω = acentric factor (Eq. 3.17)

SUBSCRIPTS

b = boiling-point property

c = critical property

C = combustion

f = formation from elements (ΔH_f) or fusion-point property (t_f)

i, j = property of component i, j

ij = characteristic property of i-j interaction

m = mixture property

o = atmospheric-pressure property (Eq. 9.3)

0 = property at $0°R$ (Eqs. 7.18a and 8.1)

r = reduced property (relative to critical value)

ref = value at reference state (e.g., 60°F)

R = reaction
V = vaporization

SUPERSCRIPTS

B = bulk property (Eq. 10.9)
L = liquid property
s = property at saturation
S = surface property (Eq. 10.9)
V = vapor property
o = standard-state property
$*$ = ideal-gas property (Ch. 6)

THERMODYNAMIC AND TRANSPORT PROPERTIES OF COAL LIQUIDS

Chapter 1

INTRODUCTION

In petroleum refining, we deal with process streams that may contain compounds ranging from hydrogen to some that have normal boiling points much higher than 1000°F. To characterize such a stream, we generally use defined compounds and undefined mixtures or fractions.

"Defined" compounds are the identifiable chemical substances: hydrogen, methane, benzene, phenol. As the carbon number of the hydrocarbon increases, however, the number of identifiable chemical substances rises rapidly. Even if only paraffins are considered, the number of isomers increases from 3 for C_5, to 5 for C_6, to 9 for C_7, to 18 for C_8, and so forth. It would be prohibitively expensive to identify all these isomers on a routine basis. It would be hopeless to try to do so for the $C_{10}+$ paraffins or for the other classes of hydrocarbons.

The introduction of "undefined" mixtures, also commonly known as "fractions," has tremendously simplified the characterization of petroleum. Defined compounds are generally used up to C_5, but all heavier compounds are grouped into "petroleum fractions." The normal practice is to break a wide petroleum cut into several narrower cuts or fractions. Each fraction should have a narrow boiling range, preferably less than 50°F, so that it can be treated as a single pseudocomponent that has only one boiling point (equal to the average boiling point of the fraction).

This approach generally is used for boiling points, t_b, above 100°F. (The normal boiling point of n-pentane, usually the heaviest identified "defined" compound, is 96.92°F; American Petroleum Institute, 1983.) A fraction boiling at 225°F, for example, would include compounds boiling in the range 200 to 250°F. Examples of such paraffins are n-heptane, 3-ethyl-pentane, and 17 C_8 isomers.

The enormous simplification introduced by the use of fractions (for $t_b > 100$°F) was very successful for paraffinic crude oils. When crudes with relatively high levels of naphthenes became important, however, the one-parameter characterization was no longer successful. A second parameter, a measure of structural differences, was needed to differentiate between naphthenes and paraffins. That was supplied by the density of the fraction.

The second parameter used in the characterization of petroleum fractions is the specific gravity at 60/60°F; that is, the density (or gravity) of the fraction at 60°F divided by the density of water, also at 60°F. Entirely equivalent to the specific gravity is the gravity expressed in °API. A third alternative is the Watson characterization factor, K_w:

$$K_w = [T_b(°R)]^{1/3}/S \tag{1.1}$$

T_b is the normal boiling point in degrees Rankine and S is the specific gravity at 60/60°F.

The introduction of S or K_w made it possible to account for naphthenes and even aromatics in the petroleum fractions. (More than 15 naphthenes would be included in the 200–250°F cut mentioned earlier.) However, most of the thermodynamic correlations are based on the properties of paraffinic/naphthenic crudes. The correction for aromatics works best when the level of aromatics is relatively small. This can be better understood by comparing values for K_w.

The aromaticity of liquids processed in refineries is commonly measured by K_w. Crude oil fractions have a K_w of about 12, while heavy paraffins have K_w's in excess of 13. The K_w of naphthenes goes down to ~11. Thus most of what is known about the properties of refinery process streams is for fractions with $K_w \gtrsim 11$.

Coal liquids are much more aromatic than petroleum crude. Coal liquids have fractions with K_w's less than 10 and even below 9. Thus any correlations that have been developed for use in refinery process calculations are either inapplicable or, at best, will need to be extrapolated into regions where essentially no experimental information is available.

The very low K_w's of coal liquid fractions are primarily due to the high levels of PNAs (polynuclear aromatic hydrocarbons) they contain. Coal liquids also contain significant amounts of heteroatoms, especially oxygen and nitrogen. In order to understand the properties of coal liquids, therefore, we must also investigate the properties of PNAs and their derivatives. Accordingly, we examine in this book two classes of compounds: coal liquids and model compounds, that is, the defined compounds that give coal liquids their characteristic properties.

The following sections compare coal liquids with petroleum crude to establish the major differences. Then the EDS coal liquids are compared with other coal liquids obtained by direct liquefaction, followed by a brief discussion on the sources of coal. Model compounds are considered more extensively, and recommendations are listed in Table 1.1. Finally, the properties and conditions of interest in this book are outlined, as an introduction to Chapter 2, and the methods used are indicated, as an introduction to Chapter 3 and later chapters.

COAL LIQUIDS COMPARED TO PETROLEUM CRUDE

Hochman (1982) compared the composition of coal liquids and other synthetic fuels with that of Arab light crude from Saudi Arabia. The coal liquids of interest are those obtained by a direct liquefaction process: SRC-I, SRC-II, H-Coal, EDS, and others. Hochman's comparison, as well as this book, focuses on the EDS coal liquids, but the following comments should apply to all "direct" coal liquids.

The two types of liquids are very different in terms of saturates (paraffins and naphthenes), olefins, and aromatics:

Chemical Type	Arab Light Crude	Direct Coal Liquids
Saturates (wt%)	80	35
Olefins (wt%)	—	5
Aromatics (wt%)	20	60

Arab light crude contains only 20 wt% aromatics, mostly alkylbenzenes ($K_w \gtrsim 10$). Direct coal liquids contain 60 wt%, and often more, aromatics, mostly PNAs ($K_w \lesssim 10$). Mass spectrometric measurements on direct coal liquids have determined that the condensed two- to five-ring compounds predominate and that about half of the rings are saturated. It is this high level of aromaticity and the presence of PNAs that raise questions about our ability to predict the thermodynamic and transport properties of coal liquids. The high content of PNAs also significantly lowers the atomic hydrogen-carbon ratio: 1.5 to 1 in direct coal liquids versus 1.9 to 1 in Arab light crude (Hochman, 1982).

The other significant difference between coal liquids and petroleum crude is the level of heteroatoms. The following comparison is again taken from Hochman (1982):

Heteroatoms	Arab Light Crude	Direct Coal Liquids
Sulfur (wt%)	1.8	0.2
Nitrogen (wt%)	0.1	0.3
Oxygen (wt%)	—	2.0
Compounds containing heteroatoms (wt%)	10	25

Arab light crude contains a high level of sulfur, very little nitrogen, and essentially no oxygen. Although EDS coal liquids contain little sulfur and nitrogen, a benefit of the internal hydrogenation step that reduces sulfur and nitrogen, they contain a very high level of oxygen, principally in phenolic compounds. Even the 0.3 wt% nitrogen in EDS coal liquids (the nitrogen content is higher in other direct coal liquids) is a problem, because it cannot be removed easily from an aromatic ring.

The presence of heteroatoms influences the properties of coal liquids, making them even more different from those of petroleum crude. As shown in the table, about 25 wt% of the compounds in coal liquids contain heteroatoms, compared to only 10 wt% in Arab light crude. In petroleum crudes, the major heteroatom is sulfur, contained in compounds that are mostly nonpolar or weakly polar. Therefore, correlations developed for hydrocarbons are more likely to apply to sulfur-containing compounds, and indeed they do. In coal liquids, however, the phenolic compounds, the major type of oxygen-containing compounds, are very polar. This is another reason the correlations for petroleum fractions are not expected to work as well for coal liquid fractions.

Heteroatoms are of special concern in hydrotreating, the process emphasized by Hochman (1982). Their presence also raises concerns about emissions during combustion. Furthermore, PNAs are toxic, and must therefore be handled with special care in combustion, wastewater treating, and other operations. Although important, these concerns are not addressed in this book.

EDS COMPARED TO OTHER DIRECT COAL LIQUIDS

In addition to EDS coal liquids, much is known about other direct coal liquids. As an example, Dooley et al. (1979) characterized coal liquids from the following processes (and coals): COED (Utah, western Kentucky), Synthoil (West Virginia), and H-Coal (Illinois No. 6). They present data on sulfur (0.05–0.42 wt%), nitrogen (0.23–0.79 wt%), distillation curve, ring-number distribution, and other properties. The highest sulfur and nitrogen levels were found in the liquid from the Synthoil process, a process no longer of interest. The liquids from the other processes are similar to those from EDS.

Hipkin et al. (1981) summarized data on "syncrudes" from SRC-I, SRC-II, H-Coal, and EDS, the most promising of the direct liquefaction processes. The heteroatom levels reported were measured by Chevron:

Heteroatom	SRC-I	SRC-II	H-Coal	EDS
Sulfur (wt%)	0.89	0.29	0.32	0.12
Nitrogen (wt%)	2.04	0.85	0.46	0.30
Oxygen (wt%)	4.52	3.72	1.80	1.92

The SRC-I syncrude has a much higher heteroatom content than the others, but is also very different in that it is solid at room temperature. Indeed, its pour point is greater than 400°F. The gravity of SRC-I is −14.6 °API ($S = 1.21$), while that of the other three syncrudes ranges from 19 to 31 °API ($S = 0.94$ to 0.87).

The characterization and properties of liquids from the SRC-II process are examined extensively in this book, because they have been investigated by J. A. Gray at Gulf Research and Development Company. The characterization and properties of the SRC-II liquid are given by Gray (1981), Gray et al. (1983), Gray and Holder (1982), and by Holder and Gray (1983).

SOURCES OF COAL

The EDS process has been used to liquefy Illinois No. 6 coal (bituminous), Wyoming Wyodak coal (sub-bituminous), and Texas lignite. Most of the data presented in this book are for the first two coals.

The SRC-II liquids examined by J. A. Gray were obtained from Pittsburgh A-seam coal (bituminous). Dooley et al. (1979) characterized liquids from a variety of coals: western Kentucky (bituminous), West Virginia (bituminous), Utah (sub-bituminous), and Illinois No. 6. The data reported by Hipkin et al. (1981) were on SRC-II liquids from Pittsburgh A-seam coal, H-Coal liquids from Illinois No. 6, and EDS liquids from Wyodak; the source of the SRC-I liquids was not identified.

Although the source of coal might be expected to affect the properties of coal liquids, it is shown in later chapters that the type of coal used did not result in significant differences in thermodynamic or transport properties. That is, a given correlation apparently predicts the properties of liquids from Illinois No. 6 coal just as well—or just as poorly—as those from Wyoming Wyodak coal. This insensitivity is probably a result, at least to some extent, of the characterization of coal liquid fractions by only boiling point and specific gravity. Such a simple characterization cannot fully account for, say, differences between alkylbenzenes and PNAs or the presence of heteroatoms. It is one of the objectives of this book to demonstrate how far one can go with the (t_b, S) characterization and the correlations based on it. Since these correlations were developed for petroleum fractions, it is also important to demonstrate whether they apply to coal-liquid fractions.

Both of these tasks can be greatly facilitated by including in the investigation model compounds (defined compounds characteristic of those found in coal liquids) and their properties.

MODEL COMPOUNDS FOR COAL LIQUIDS

Model compounds are especially important in developing physical models and correlations for characterizing and predicting properties of coal liquids. For example, data for the vapor pressure of model compounds can be used to test or develop vapor pressure correlations for coal-liquid fractions.

In addition to this indirect use, there are at least three direct uses of model compounds. One is when a defined compound is present in the process streams in substantial amounts, such that its properties must be known accurately. An example is tetralin, an important compound in the EDS coal liquefaction process.

Another direct use of model compounds arises in the investigation of *future* processes. Such investigations, especially in their preliminary phases, are often based entirely on defined compounds. For example, in the early stages of the EDS process development, there was interest in the hydrogenation of naphthalene to tetralin.

Finally, model compounds are important in the study of highly complex reactions. An example of such reactions is denitrogenation, an essential step in the production of stable synthetic fuels. Because denitrogenation of coal liquids is not yet well understood, current investigations focus on simple model compounds like quinoline and carbazole.

As discussed, the major characteristics of coal liquids that distinguish them from petroleum liquids are their aromaticity and high heteroatom content. The aromatic hydrocarbons of special interest are PNAs with two to five rings and, especially, their partially hydrogenated derivatives. The side chains are generally short: C_1 to C_3. The most important of the heteroatoms is oxygen, which is present primarily in phenolic compounds. Also extensively present is nitrogen, especially in the unhydrotreated coal liquids, while the least important heteroatom is sulfur.

In addition to aromaticity and heteroatom content, two other conditions were considered in selecting model compounds: that their normal boiling point be above 400°F; and that some of their properties be known and, whenever possible, the properties of their partially hydrogenated derivatives. The boiling point condition was imposed to focus attention on the heavier compounds, for which relatively little is known and for which the available correlations may not be applicable.

The list of model compounds for coal liquids in Table 1.1 is certainly neither exhaustive nor inviolate. Hundreds of compounds could be added or compounds on the list could be replaced by similar ones; for example, 1-ethylnaphthalene can replace 1-methylnaphthalene. However, it is be-

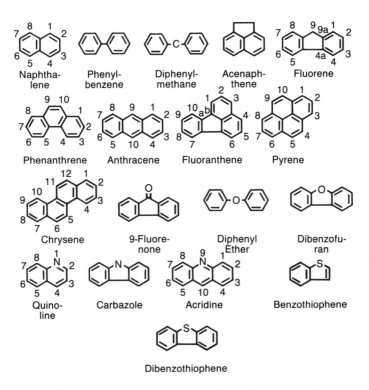

Figure 1.1 Structure of selected model compounds.

lieved that the compounds listed represent the major characteristics of the coal liquids that distinguish them from petroleum liquids. Measurements on these model compounds can provide the "anchor" points on which to base new correlations for the properties of aromatic compounds and coal liquids.

Table 1.1 presents the molecular weight, boiling point, specific gravity, and Watson characterization factor of the model compounds. These data are used in Chapter 3 and the following chapters. The freezing point, t_f, is also included to identify those compounds that are solid at 60°F. The structure of selected unhydrogenated compounds is given in Figure 1.1.

Having introduced the coal liquids and appropriate model compounds, we also need to indicate what thermodynamic and transport properties and what conditions are of primary concern in this book.

Table 1.1   Molecular Weight, Normal Boiling and Freezing Points, Specific Gravity at 60/60°F, and Watson Characterization Factor of Model Compounds for Coal Liquids

Compound	M^a	t_b (°F)	t_f (°F)	S^b	$K_w{}^c$
Naphthalene	128.173	424.39^d	176.51^d	$1.030^{d,e}$	9.318
Tetralin (1,2,3,4-tetrahydronaphthalene)	132.205	405.72^f	-32.35^f	0.9739^f	9.785
cis-Decalin (cis-Decahydronaphthalene)	138.252	384.47^g	-45.36^g	0.9011^g	10.488
1-Methylnaphthalene	142.200	472.43^h	-22.86^h	1.0244^h	9.536
2-Methylnaphthalene	142.200	465.89^h	94.24^h	$1.000^{e,h}$	9.745
Phenylbenzene	154.211	491.0^h	156.6^h	$1.027^{e,h}$	9.574
Phenylcyclohexane	160.258	460.6^i	44.73^i	$0.947^{i,j}$	10.27
Cyclohexylcyclohexane	166.306	455.0^i	38.61^i	$0.889^{i,j}$	10.92
Diphenylmethane	168.238	507.69^h	77.43^h	$1.0104^{e,h}$	9.788
Acenaphthene	154.211	531.30^k	200.14^k	$1.094^{e,k}$	9.11
Fluorene	166.222	567.32^k	238.63^k	$1.09^{e,l}$	9.26
1,2,3,4,4a,9a-Hexahydrofluorene	172.269	512.6^i	44 ± 1^i	$0.992^{i,j}$	9.99
Phenanthrene	178.233	644.5^m	210.47^m	$1.130^{e,m}$	9.15
9,10-Dihydrophenanthrene	180.249	597.4^o	92.07^o	$1.09^{e,l}$	9.35
1,2,3,4-Tetrahydrophenanthrene	182.265	$(590)^p$	91.4^h	$1.079^{e,h,j}$	9.42
1,2,3,4,5,6,7,8-Octahydrophenanthrene	186.296	$563.^h$	62.6^h	$1.0281^{e,h}$	9.80
Anthracene	178.233	647.65^m	421.0^m	$1.123^{e,m}$	9.21
9,10-Dihydroanthracene	180.249	$581^h\text{–}599^i$	226.4^h	$1.08^{e,l}$	9.38–9.44
1,2,3,4-Tetrahydroanthracene	182.265	$590.^h$	219.2^h	$1.06^{e,l}$	9.59
1,2,3,4,5,6,7,8-Octahydroanthracene	186.296	561.2^h	162.03^q	$1.02^{e,h,j}$	9.87
Fluoranthene	202.255	721.0^k	230.32^k	$1.17^{e,l}$	9.03
1,2,3,10b-Tetrahydrofluoranthene	206.287	$689.^i$	$167.^i$	$1.11^{e,i,j}$	9.43
Pyrene	202.255	742.6^r	302.4^r	$1.19^{e,j,r}$	8.94
Chrysene	228.293	825.8^r	496.4^r	$1.20^{e,l}$	9.06

m-Cresol (3-methyl-1-hydroxybenzene)	108.140	396.10[s]	54.00[s]	1.0385[s]	9.14
2,4-Xylenol (2,4-dimethyl-1-hydroxybenzene)	122.166	411.76[t]	76.15[t]	1.0248[e,t]	9.32
1-Naphthol (1-hydroxynaphthalene)	144.173	551.1[i]	205.[i]	1.16[e,i,j]	8.65
5,6,7,8-Tetrahydro-1-naphthol	148.204	515[i]	156.[i]	1.08[e,l]	9.18
9-Hydroxyfluorene	182.221	665.[u]	—	1.20[l]	8.67
2-Phenanthrol (2-hydroxyphenanthrene)	194.232	743.[i]	336.[i]	1.23[e,l]	8.65
1,3-Dihydroxybenzene (resorcinol)	110.112	529.7[i]	231±1[i]	1.28[e,l]	7.78
9-Fluorenone	180.206	646.66[u]	—	1.21[l]	8.55
Diphenyl ether	170.210	496.7[i]	80.4[i]	1.077[e,i]	9.15
Dibenzofuran	168.195	543.97[v]	180.5[v]	1.16[e,v]	8.63
Quinoline	129.161	459.7[w]	5.18[w]	1.0986[w]	8.851
1,2,3,4-Tetrahydroquinoline	133.193	484.[i]	68.[i]	1.062[e,i,j]	9.24
Carbazole	167.210	670.57[x]	475.0[x]	1.178[e,x]	8.84
Acridine	179.221	654.8[x]	234.[x]	1.11[e,x]	9.34
1,2,3,4-Tetrahydroacridine	183.252	622.9[i]	133.[i]	1.13[e,l]	9.09
Benzothiophene	134.195	427.8[i]	88.4[i]	1.11[e,l]	8.66
Dibenzothiophene	184.255	629.6[i]	210[i]	1.19[e,l]	8.65

[a]IUPAC (1979); [b]Density of compound at 60°F divided by the density of water at 60°F (0.999024 g · cm⁻³), both measured in air at 14.696 psia, but corrected to vacuum; [c]$K_w = [T_b(°R)]^{1/3}/S$; [d]Kudchadker et al. (1978c); [e]For supercooled liquid (extrapolated); [f]Kudchadker et al. (1978b); [g]Chaback (1978); [h]TRCHP Tables (1983); [i]Anderson and Wu (1963); [j]Extrapolated from density data at higher temperatures with approximate values for $\Delta\rho/\Delta t$ (Gray, 1984); [k]Kudchadker et al. (1981a); [l]Estimated from t_b and M (Gray, 1984); [m]Kudchadker et al. (1979a); [n]Wieczorek and Kobayashi (1981); 602.6°F in h; [o]Lee-Bechtold et al. (1979); 93.2°F in h; [p]Approximate; should be between the t_b's of the di- and octahydrophenanthrenes, and close to the t_b of 1,2,3,4-tetrahydroanthracene; [q]Gammon et al. (1982); 163.4°F in h; [r]Kudchadker et al. (1979b); [s]Kudchadker et al. (1978a); [t]Kudchadker and Kudchadker (1978); [u]Sivaraman et al. (1983); [v]Wilhoit and Adler (1983); [w]Viswanath (1979); [x]Kudchadker et al. (1981b).

PROPERTIES AND CONDITIONS OF INTEREST

As already noted, we are emphasizing those characteristics of coal liquids that make them different from petroleum liquids; for example, the high level of PNAs and oxygen. In addition, we are emphasizing properties at high temperatures. The liquefaction reactor operates at about 850°F for the direct liquefaction processes SRC-I, SRC-II, H-Coal, and EDS. Very little is known about properties at such high temperatures, even for petroleum fractions. The liquefaction reactor pressure is about 2000 psia for SRC-I, SRC-II, and once-through EDS; 2500 psia for vacuum bottoms-recycle EDS; and 3000 psia for H-Coal. Also of major concern, as discussed in Chapter 2, is the vacuum pipestill, where the temperature may be close to 700°F—and the pressure perhaps only 25 mm Hg (0.5 psia). Thus, the temperature range of primary interest is 700 to 850°F. We are also concerned with how the high pressure, largely due to the hydrogen and, to a lesser extent, methane partial pressure, affects the properties of the coal liquids.

The effluent of a coal liquefaction reactor contains, in addition to liquids, solids (unconverted coal and ash) and gases (hydrogen, methane, and others). In this book we focus our attention on the properties of the *liquids*, mostly those boiling between 400 and 1000°F. Gases, especially hydrogen and methane, are included in our examination of vapor-liquid equilibria and surface tension. They also are examined insofar as their effect on liquid properties. However, with the exception of ideal gas heat capacity, no data or correlations for gas properties are presented. The level of PNAs in the gas phase should be very small, and thus the gas mixtures, as well as their properties, should not differ greatly from those found in petroleum refinery process streams.

Solids are much more important than gases and are peculiar to coal liquefaction. Solids, in particular, coal and ash, generally are not present in petroleum refinery process streams. Owing to the importance of the solids, special procedures have been developed for calculating their properties; see, for example, the DOE/IGT *Coal Conversion Systems Technical Data Book* (1984).

Although we do not consider the properties of solids, in Chapter 2 we comment on how the properties of solids and liquids can be blended to predict the properties of coal/liquid slurries, especially those needed in heat and material balance calculations, that is, heat capacity and density. Another property of special importance in coal liquefaction is the viscosity of the coal/liquid slurry.

METHODS

Finally, a few words are needed to describe the methods we use in Chapters 3 through 9. As noted in the Preface, our concern is the methods—the characterization and property correlations—used in the *industry* for *petroleum* liquids. Since these are the methods that most design engineers are now using, we need to demonstrate any limitations when they are applied to coal liquids. We consider only limited modifications of these methods, as needed to improve the predictions for coal liquids. The basis for two broad types of methods we use is given in Chapter 3.

Methods not presently used in the industry, current research on characterization and property prediction for heavy liquids, and future data needs are examined in Chapter 10.

REFERENCES

American Petroleum Institute, *Technical Data Book—Petroleum Refining*, 4th Ed., API, Washington, DC, 1983.

Anderson, H., and W. R. K. Wu, *Properties of Compounds in Coal-Carbonization Processes*, Bureau of Mines Bulletin 606, U.S. Dept. of the Int., Washington, DC, 1963.

Chaback, J. J., *cis- and trans-Decalin*, API Publication 706, 1978.

Department of Energy/Institute of Gas Technology, *Coal Conversion Systems Technical Data Book*, Chicago, IL, March 1984.

Dooley, J. E., W. C. Lanning, and C. J. Thompson, Characterization Data for Syncrudes and Their Implication for Refining, in M. L. Gorbaty and B. M. Harney (eds.), *Refining of Synthetic Crudes*, Advances in Chemistry Series No. 179, American Chemical Society, Washington, DC, 1979, pp. 1–12.

Gammon, B. E., et al., Heat Capacity, Vapor Pressure, and Derived Thermodynamic Properties of Octahydroanthracene, *Proc. 8th Sympos. Thermophys. Properties*, Vol. II, ASME, New York, 1982, p. 402.

Gray, J. A., Selected Physical, Chemical, and Thermodynamic Properties of Narrow Boiling Range Coal Liquids from the SRC-II Process, Report No. DOE/ET/10104-7, April 1981.

Gray, J. A., C. J. Brady, J. R. Cunningham, J. R. Freeman, and G. M. Wilson, Thermophysical Properties of Coal Liquids. 1. Selected Physical, Chemical, and Thermodynamic Properties of Narrow Boiling Range Coal Liquids, *Ind. Eng. Chem. Process Des. Dev.*, **22**, 410 (1983).

Gray, J. A., and G. D. Holder, Selected Physical, Chemical, and Thermodynamic Properties of Narrow Boiling Range Coal Liquids from the SRC-II Process, Supplementary Property Data, Report No. DOE/ET/10104-44, April 1982.

Gray, R. D., Jr., personal communication (1984).

Hipkin, H. G., T. C. Lin, and K. C. Lu, *Survey of Design Data Needs for Refining Synthetic Crudes*, American Petroleum Institute, Washington, DC, December 1981.

Hochman, J. M., Synthetic Fuels: Are They Different? *CHEMTECH*, **12**, 500 (1982).

Holder, G. D., and J. A. Gray, Thermophysical Properties of Coal Liquids. 2. Correlating Coal Liquid Densities, *Ind. Eng. Chem. Process Des. Dev.*, **22**, 424 (1983).

IUPAC, Atomic Weights of the Elements 1977, *Pure Appl. Chem.*, **51**, 405 (1979).

Kudchadker, A. P., and S. A. Kudchadker, *Xylenols*, Key Chemicals Data Books, Texas A & M U., College Station, 1978.

Kudchadker, A. P., S. A. Kudchadker, and R. C. Wilhoit, *Cresols*, Key Chemicals Data Books, Texas A & M U., College Station, 1978a.

————, *Tetralin*, API Publication 705, 1978b.

————, *Naphthalene*, API Publication 707, 1978c.

————, *Anthracene and Phenanthrene*, API Publication 708, 1979a.

————, *Four-Ring Aromatic Condensed Compounds*, API Publication 709, 1979b.

Kudchadker, A. P., S. A. Kudchadker, R. C. Wilhoit, and S. K. Gupta, *Acenaphthylene, Acenaphthene, Fluorene, and Fluoranthene*, API Publication 715, 1981a.

————, *Carbazole, 9-Methylcarbazole, and Acridine*, API Publication 716, 1981b.

Lee-Bechtold, S. H., et al., A Comprehensive Thermodynamic Study of 9,10-Dihydrophenanthrene, *J. Chem. Thermodynamics*, **11**, 469 (1979).

Sivaraman, A., R. J. Martin, and R. Kobayashi, A Versatile Apparatus to Study the Vapor Pressure and Heats of Vaporization of Carbazole, 9-Fluorenone and 9-Hydroxyfluorene at Elevated Temperatures, *Fluid Phase Equilibria*, **12**, 175 (1983).

TRCHP (Thermodynamic Research Center Hydrocarbon Project), *Selected Values of Properties of Hydrocarbons and Related Compounds*, Texas A & M U., College Station, sheets extant 1983.

Viswanath, D. S., *Quinoline*, API Publication 711, 1979.

Wieczorek, S. A., and R. Kobayashi, Vapor-Pressure Measurements of 1-Methylnaphthalene, 2-Methylnaphthalene, and 9,10-Dihydrophenanthrene at Elevated Temperatures, *J. Chem. Eng. Data*, **26**, 8 (1981).

Wilhoit, R. C., and S. Adler, *Benzofuran, Dibenzofuran and Benzonaphthofurans*, API Publication 721, 1983.

Chapter 2

BASIC DATA NEEDS
FOR DESIGN

Much has been written about data needs for design and how uncertainties in key properties impact on the size or even the operability of a plant. Here we summarize a few general observations and "truths," and then we focus on the key data issues in coal liquefaction. Much of the information we present comes from a workshop on coal liquids (Brinkman and Reilly, 1981). Finally, we consider key data issues in the EDS process, many of which are common to all direct coal liquefaction processes.

GENERAL CONSIDERATIONS

The first and most important use of thermodynamic properties is in heat and material balance calculations. In such calculations, any uncertainties in the data or in the predictions will directly affect the results. The three essential properties are:

Vapor–liquid equilibria

Enthalpy

Density or, more generally, P-V-T (pressure-volume-temperature) properties

Vapor–liquid equilibria, which include the vapor pressure of the defined compounds or fractions, are particularly important because most separations are effected through partial vaporization of a liquid mixture. (If separation is by liquid–liquid extraction, then liquid–liquid equilibria become important.) Predicting vaporization is difficult in multicomponent systems, especially those that are nonideal. According to Nagel et al. (1980), nearly 60% of the BASF budget on "basic" data goes to measuring vapor pressures and VLE (vapor–liquid equilibria). (Another 4% goes to measuring liquid–liquid equilibria.) About 11% is spent on caloric data, that is, heat capacity, heat of vaporization, heat of mixing, and, what is often the most important, heat of reaction.

BASF also is interested in measuring density and viscosity, but Nagel et al. give no breakdown of the BASF budget for these properties. They also note that 1% of the basic data budget goes for thermal conductivity measurements (and 27% for reaction kinetics, which is of no concern here).

Although the BASF budget allocation may be more typical of chemical companies, the interests of petroleum companies are similar. VLE and enthalpy *are* the top-priority properties.

Transport properties and surface tension are of lesser interest because they are primarily used in the design of equipment. Nagel et al. (1980) summarized the reasons why great accuracy might not be needed in the properties used for equipment design:

1. The property may have little influence on equipment size because its value may not vary greatly or it is frequently raised to a fractional exponent.

2. The correctness of a model used in equipment design may be more important than accuracy of the data. (It may be more correct to say that, if a model has been based on lower-quality data, its use with high-quality data may make predictions worse.)

3. Fouling of equipment cannot be quantified; therefore, the size of equipment is less affected by inaccuracies in thermodynamic and transport property data.

Nagel et al. also discussed how various properties enter in the design of distillation towers, liquid–liquid extractors, heat exchangers, gas–liquid reactors, and other process equipment.

Najjar et al. (1981) demonstrated how heat transfer coefficients are affected by uncertainties in physical property data. They considered uncertainties in the liquid density, viscosity, thermal conductivity, and heat capacity. A 50% error in each of these properties resulted in a calculated heat transfer coefficient that was from 60% low to 110% high. (A 50% error is unrealistic for liquid density and even for heat capacity, but not too rare for viscosity and thermal conductivity.)

Mersmann (1980) reviewed data needs in heat and mass transfer. Many more studies of data needs have been reported, but most of them concern cryogenic processes. The following two sections focus on data needs specific to coal liquefaction.

KEY DATA ISSUES IN COAL LIQUEFACTION

As noted, VLE and enthalpies are required for heat and material balance calculations, the first step in process design. Reliable data and models are needed to predict the extent of vaporization and the distribution of the key components between the vapor and liquid phases. In some cases, there may be more than one liquid phase, for example, an organic-rich and water-rich phase, and even solid phase(s). Indeed, the presence of solids, unconverted coal and ash, in coal liquefaction process streams significantly complicates the calculation of thermophysical properties. Although these solids do not affect the VLE behavior, they have a large effect on other properties; for example, slurries are non-Newtonian and, as such, their viscosity is a function of shear rate and generally cannot be predicted accurately.

Viscosity and the other transport properties, thermal conductivity and diffusivity, as well as surface tension, are important in equipment design. For reasons that have already been discussed, these properties generally are less important than VLE and enthalpy, although the viscosity of slurries is a key property. Of intermediate importance are the P-V-T properties, especially the density of liquids and dense fluids, which are needed both in material balance and equipment design calculations.

Temperature/enthalpy curves are needed very early in process design to establish heat loads and to optimize the process heat integration. If a reaction is involved, the temperature/enthalpy curve must also reflect the contribution of the heat of reaction. This is a thermochemical property that can be calculated accurately if the heat of formation (or the equivalent heat of combustion) is known for each process stream component involved in the reaction. Although heats of formation are known for a wide variety of defined compounds, they generally have to be estimated for fractions. The heat of combustion of petroleum fractions can be predicted with reasonable accuracy (American Petroleum Institute, 1983), but this has not yet been established for coal liquids.

Along with the heat of formation, we need the Gibbs free energy of formation to predict the equilibrium distribution of the reactants and the reaction products. No reliable correlations exist for the Gibbs free energy of formation of any fraction, let alone coal-liquid fractions. We can do better with defined compounds, and that can prove very useful in developing third-generation processes, that is, processes that are not yet investigated even in the laboratory. Such development can be based on model compounds, as noted in Chapter 1. With data for the thermochemical properties of the model compounds, it would be possible to investigate, first, whether a proposed liquefaction reaction is feasible, and then to establish the optimum temperature and pressure for the reaction.

Thermochemical properties (heats and Gibbs free energies of formation) for model compounds would also be useful in investigating a difficult problem such as the aging of coal liquids. Aging and the related problem of denitrogenation of coal liquids involve a combination of chemical and physical effects. It would be hopeless to try to understand what goes on in an actual coal liquid when we may not yet know, for example, how quinoline is denitrogenated. And, then, we would also like to know how to effect this denitrogenation with minimum hydrogen consumption and energy input.

Pure Compounds and Mixtures

In the 1981 workshop on "Design Properties of Coal Liquids" (Brinkman and Reilly, 1981), the working group on pure compounds focused on the following properties, in approximate order of priority:

Critical properties
Vapor pressure of liquid
Density of saturated liquid (and vapor)
Enthalpy of liquid and vapor
Viscosity of liquid (and vapor)
Vapor pressure and heat capacity of solid
Surface tension
Thermal conductivity of liquid (and solid and vapor)
Heat and Gibbs free energy of formation

The emphasis on critical properties is due to the importance of corresponding-states methods in the prediction of properties; see Chapter 3 and the later chapters. The properties given in parentheses were considered of lesser importance.

What should be emphasized in this list is that it refers to current direct liquefaction processes; that is, SRC-I, SRC-II, H-Coal, EDS. When *future* processes are considered, the heat and Gibbs free energy of formation become much more important.

In the same workshop, the working group on binaries and multicomponent mixtures, not unexpectedly, found that the most pressing need was for phase equilibrium and enthalpy data. Such data are used, as already noted, in heat and material balance calculations and in the design of separators, heat exchangers, and other equipment. Prediction of transport properties, needed only in equipment design, was given a lower priority.

Critical Design Areas

Equipment design and operability were the focus of the working group on design engineering. The critical design areas were determined to be, in order of priority:

Furnaces
Heat exchangers
Phase separators
Letdown system and vacuum flash
Hydrogenation

No mention was made of the liquefaction reaction.

For the vacuum flash, or the vacuum pipestill, the pumpability of the bottoms product is a key concern. This stream may contain 50 wt% solids and has a very high viscosity; more on this subject is included in the next section.

This same group noted the potential problem of ammonium salt deposition in heat exchangers and the need for water injection to remove such deposits. Ammonium salt deposition was also mentioned by the working group on slurries as a possible cause of corrosion, especially in fractionators. (A similar problem exists in petroleum refinery fractionators.)

In view of the presence of NH_3 in process streams, together with acidic gases such as H_2S, HCl, and CO_2, an engineer certainly must evaluate the potential for formation of ammonium salts. In the absence of liquid water, such salts will deposit on equipment, leading to plugging and the concomitant corrosion. Although water injection will generally dissolve these salts and thus eliminate the problem, it is preferable that the formation of ammonium salts be *prevented*. This can be accomplished by reducing the level of NH_3 or of the acidic gas(es); sometimes, it may only be necessary to modify the design so that the temperature is higher than that favoring the formation of solid ammonium salt(s).

Another multiphase equilibrium problem is the water/coal liquid mutual solubility. As noted in a recent paper (Tsonopoulos and Wilson, 1983), the formation of a second, water-rich, liquid phase may be intentional—to remove, for example, the ammonium salts—or may be unwanted with harmful effects such as corrosion and poor product quality. The solubility of coal liquid fractions (excluding those rich in phenols) in water is very low and is primarily of concern in water pollution abatement, a topic not considered in this book. On the other hand, the solubility of water in coal liquids (and in petroleum liquids) can be very high, especially at elevated temperatures, and thus may need to be taken into account, together with the volatility of water, in designing distillation towers and other equipment.

KEY DATA ISSUES IN THE EDS PROCESS

Although this section emphasizes the EDS process, the issues are similar for SRC-I, SRC-II, and H-Coal.

The principal areas of uncertainty in thermodynamic and transport property data, as they may affect the design of an EDS coal liquefaction plant, are:

1. VLE at the coal slurry furnace outlet/liquefaction reactor inlet conditions (~850°F and 2000 psia, for once-through operation, to 2500 psia, for vacuum-bottoms recycle operation).
2. VLE at the vacuum pipestill operating conditions (~700°F and 25 mm Hg; feedstream containing ~25 wt% solids).
3. Viscosity/pumpability of the vacuum pipestill bottoms slurry (1000°F+ material with as much as 50 wt% solids).
4. Viscosity of liquefaction feed slurry in the slurry preheat furnace (~30 to 45 wt% solids in hydrogenated solvent).

Item 1 is important because it determines the operability of the reactor. Overprediction of vaporization leads to a conservative design. However, if vaporization is excessive, the reactor would be inoperable. Vaporization in the liquefaction reactor effluent may strongly affect the recoverability of materials boiling below 1000°F from the 1000°F+ residuum. The temperature and pressure conditions are very similar to those in SRC-I and -II (~850°F and 2000 psia), as well as H-Coal (~850°F and 3000 psia).

The H_2 partial pressure is about 80% of the total pressure and that of methane is about 10%. The volatility and solubility of these components in the coal liquid are key VLE needs. The volatility of H_2S, which is produced in the liquefaction reactor, is important in determining the size of the H_2S-removal facilities downstream. Also important is the volatility of H_2O and NH_3.

The vacuum pipestill treats the bottoms product of the atmospheric pipestill, which operates at ~700°F and 50–100 psia. The bottoms product of the vacuum pipestill is about 30–35 wt% of the dry coal feed (bottoms recycle operation). SRC-II and H-Coal use a vacuum flash for the final liquids/solids separation, while SRC-I uses filtration.

Items 1 and 2 underline the need for high-temperature data. VLE, or any other data, are very rare at 700–850°F, even for petroleum crude fractions. Thus experimental measurements need to be made at these high temperatures from about 3000 psia (largely due to H_2 and methane) to vacuum (where little or no H_2 or methane are present).

Items 3 and 4 relate to the properties of slurries, which deserve a brief mention here. Slurries are of special concern in all direct coal liquefaction processes. As already mentioned, the bottoms product of the vacuum pipestill can contain about 50 wt% solids. Such a stream is very viscous, possibly up to about 50 P, and is a concern for pumping.

PROPERTIES OF SLURRIES

The viscosity of slurries is probably their most important property. Slurries are non-Newtonian, and therefore it is generally impossible to predict their viscosity accurately. Fortunately, several experimental investigations of the rheology of slurries have been published (Droege et al., 1982; Florez, 1981; Lee et al., 1983; Oswald et al., 1980; Rodgers et al., 1981; Thurgood et al., 1982) that contain information on the viscosity and density of coal slurries, as well as on the swelling of coal.

Two slurry properties of interest in heat and material balance calculations are heat capacity (Droege et al., 1982; Mehta et al., 1982) and density (Droege et al., 1982; Lee et al., 1983). If no swelling occurs, both of these properties can be adequately predicted by blending the solids and liquid properties on a weight and volume basis, respectively. For example, the predicted heat capacity of a slurry that has 38.5 wt% coal (with a heat capacity of 0.43 cal/g · °C) and 61.5 wt% solvent (with a heat capacity of 0.61 cal/g · °C) is

$$(0.43)(0.385) + (0.61)(0.615) = 0.54 \text{ cal/g} \cdot °C$$

in agreement with the the result of Mehta et al. (1982).

Since the solids, mostly unconverted coal and ash, do not affect the vaporization of the liquid, the VLE (and surface tension) calculations should be made on the solids-free stream, and then the solids should be added to the resulting liquid.

A final comment on slurries concerns their reactivity. The "unconverted" coal in a slurry may continue to react to form liquid products; thus the constitution of slurries may be a function of time as well as of temperature, making it necessary to measure their properties "on-line." (Such on-line measurements may also be necessary for liquids that become solid at room temperature; for example, streams containing nonhydrotreated 700°F+ material.) On-line measurement of the viscosity of slurries, at process conditions, may be needed for the design of the coal slurry preheater, pumps, and the vacuum pipestill furnace and transfer line. On-line measurement of density is of less concern.

The thermodynamic and transport properties of coal liquids are examined in Chapters 4 through 9. To predict these properties, the coal liquids must first be adequately characterized. Chapter 3 presents such information for coal liquid fractions and model compounds.

REFERENCES

American Petroleum Institute, *Technical Data Book—Petroleum Refining*, 4th Ed., API, Washington, DC, 1983.

Brinkman, D. W., and M. J. Reilly, Design Properties of Coal Liquids: Edited Workshop Proceedings, Fountainhead Lodge, OK, March 2–4, 1981; U.S. Department of Energy, CONF-810381, August 1981.

Droege, J. W., G. H. Strickford, J. R. Longanbach, R. Venkateswar, and S. P. Chauhan, Final Report on Thermophysical Properties of Coal Liquids, U.S. Department of Energy, BMI No. 2092, WC-90d, April 23, 1982.

Florez, M. R., Rheology of EDS Process Coal Liquefaction Streams, presented at the 1981 Annual AIChE Meeting, New Orleans, November 8–12, 1981.

Knapp, H., and S. I. Sandler, *Proceedings Part II—Manuscripts of Invited Papers, Phase Equilibria and Fluid Properties in the Chemical Industry*, 2nd International Conference, W. Berlin, March 17–21, 1980. (See references for Mersmann and Nagel.)

Lee, D. D., M. R. Gibson, T. L. Sams, and J. H. Wilson, Rheology of Coal Slurries at Process Temperatures and Pressures, presented at the 1983 National AIChE Meeting, Houston, March 27–31, 1983.

Mehta, D. C., S. C. Weiner, J. R. Freeman, and G. M. Wilson, Heat Capacity of Slurry/Hydrogen Mixtures, presented at the 1982 Annual AIChE Meeting, Los Angeles, November 14–18, 1982.

Mersmann, A. B., Review of Data Needs in Heat and Mass Transfer, in Knapp and Sandler (1980), pp. 557–572.

Nagel, O., M. Molzahn, and G. Wickenhäuser, Phase Equilibria and Fluid Properties: General View by the User in Chemical Industry, in Knapp and Sandler (1980), pp. 169–204.

Najjar, M. S., K. J. Bell, and R. N. Maddox, The Infleunce of Improved Physical Property Data on Calculated Heat Transfer Coefficients, *Heat Transfer Eng.*, **2**(3/4), 27 (1981).

Oswald, G. E., E. L. Youngblood, J. R. Hightower, Jr., and J. R. Thurgood, Rheological Characterization of Coal-Solvent Slurry at High Temperature and Pressure, presented at the 1980 National AIChE Meeting, Philadelphia, June 8–12, 1980.

Rodgers, B. R., J. K. Johnson, D. D. Lee, J. H. Wilson, E. L. Youngblood, and J. R. Hightower, Experimental Support for Coal Conversion Demonstration Projects at the Oak Ridge National Laboratory: Preheater Rheology, Slurry Mixing, and Vacuum Bottoms Viscosity, presented at the 1981 Annual AIChE Meeting, New Orleans, November 8–12, 1981.

Thurgood, J. R., R. W. Hanks, G. E. Oswald, and E. L. Youngblood, The Rheological Characterization of Coal Liquefaction Preheater Slurries, *AIChE J.*, **28**, 111 (1982).

Tsonopoulos, C., and G. M. Wilson, High-Temperature Mutual Solubilities of Hydrocarbons and Water. I. Benzene, Cyclohexane, and *n*-Hexane, *AIChE J.*, **29**, 990 (1983).

Chapter 3

CHARACTERIZATION OF COAL LIQUIDS AND MODEL COMPOUNDS

Two types of correlations are used in predicting the properties of coal liquids. One is the "petroleum"-fraction correlations; these generally are based on the normal boiling point and the specific gravity at 60/60°F of the fraction. The other is the defined-compound correlations; these generally are based on the critical constants and the acentric factor of the defined compound. To use the defined-compound correlations with coal liquid fractions, the critical constants and the acentric factor of the fractions must first be determined so that the fractions can then be represented as pseudo-components.

The characterizations of fractions and model compounds, that is, defined compounds characteristic of those found in coal liquids, are examined separately in the following sections. In each case, recommendations are based on comparing several correlations with available data.

CHARACTERIZATION OF FRACTIONS

As noted by Maxwell (1950), the average specific gravity is a property of a (petroleum) fraction that can be measured directly. However, because a fraction is a mixture of a large number of components, its average normal boiling point cannot be measured. Various average boiling points can be determined, and it has been found empirically that the use of appropriate averages can lead to a better correlation of a given property; Maxwell's suggestions for which averages correlate best with specific properties are given in Table 3.1.

The volume (or weight) average boiling point can readily be calculated from the distillation curve, which is temperature versus liquid volume (or weight) percent distilled. For narrow-boiling fractions, say, $\Delta t_b \lesssim 50°F$, all averages approach each other. Therefore, the volume (or weight) average boiling point can be used to represent all of them.

Table 3.1   Average Boiling Points Used in
Property Correlations[a]

Average Boiling Point	Property
Mean average	Specific gravity Watson characterization factor Molecular weight Pseudocritical pressure Heat of combustion
Molar average	Pseudocritical temperature Thermal expansion
Weight average	True critical temperature
Volume average	Liquid heat capacity Viscosity

[a]From Maxwell, 1950.

The various average boiling points of a wide cut can be determined by appropriate blending of the volume average boiling points of the narrow fractions. However, MeABP (mean average boiling point) is defined as the boiling point that best correlates the molecular weight of (petroleum) fractions (Maxwell, 1950). A direct calculation of MeABP is given in the API Technical Data Book (1983):

$$\text{MeABP (°R)} = \frac{1}{2}\left[\sum_{i=1}^{n} x_i T_{bi} + \left(\sum_{i=1}^{n} x_{vi} T_{bi}^{1/3} \right)^3 \right] \tag{3.1}$$

x_i is the mole fraction; x_{vi}, the volume fraction of component i; and T_{bi}, the normal boiling point of component i in degrees Rankine.

Distillation Curve: Boiling-Point Distribution

Distillation curves for EDS coal liquids were determined by temperature-programmed gas chromatography. The method used is similar to that reported by Green (1976) for analyzing heavy hydrocarbon fractions. Application of this method to SRC-II coal liquids was recently reported by Pannell and Sood (1982).

The first step in GCD (gas chromatographic distillation) is to obtain a calibration curve so that the retention times of the various peaks can be related to their boiling points. To do this, it is necessary to know the level and type of aromatics present in coal liquids because aromatic compounds have shorter retention times than paraffinic compounds.

As discussed in Chapter 1, coal liquids are made up primarily of condensed multiring compounds. It was determined by mass spectrometry that the condensed two- to five-ring compounds predominate and that about half the rings are saturated. For this reason, the estimated calibration curve is intermediate between the aliphatic and aromatic curves shown in Figure 3.1. It stays close to the aliphatic curve to about 400°F, where it separates and follows a path about halfway between the aliphatic and aromatic curves. The calibration at 700°F could differ from the true boiling point by as much as 30°F; at 1000°F, probably close to the maximum temperature at which GCD currently can be relied on, the uncertainty may be 50°F.

More work is clearly needed on the characterization of coal liquids, especially those boiling near and above 1000°F. Future needs in characterization are discussed in Chapter 10.

The GC distillations of three EDS Illinois No. 6 coal liquids are plotted in Figure 3.2. GCD is assumed to be equivalent to the true boiling point analytical distillation, within the limitations just mentioned. Coal liquids such as those shown in Figure 3.2 are generally represented by 10 or more

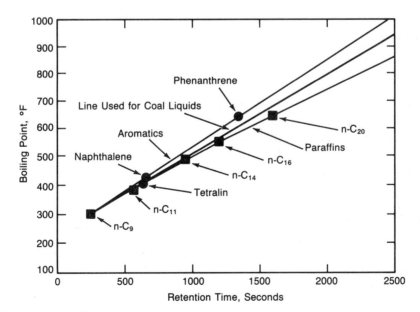

Figure 3.1   Boiling point of hydrocarbon versus retention time of gas chromatograph sampler.

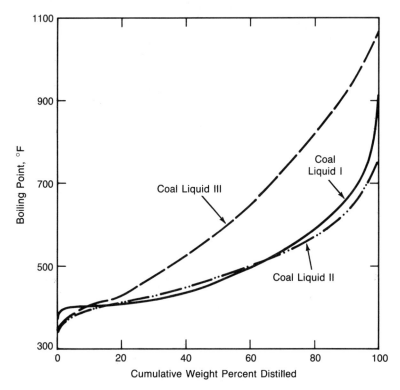

Figure 3.2 Gas chromatographic distillation of three EDS Illinois No. 6 coal liquids (see Chapter 5).

fractions. This is accomplished by breaking the smoothed distillation curve into 10 (or more) narrow-boiling ($\Delta t_b \leq 50°F$) fractions. Each fraction is identified by its average boiling point and specific gravity. This second characterization parameter, which introduces the effect of structure, is discussed next.

Specific Gravity

The minimum information needed to characterize a cut is its distillation curve and S, its average specific gravity at 60/60°F. Sometimes, the gravity of the cut (or the fraction) may be reported only in °API. In that case, Eq. 3.2 should be used to calculate S:

$$S = \frac{141.5}{°API + 131.5} \tag{3.2}$$

If only T_b and K_w are known, then

$$S = [T_b(°R)]^{1/3}/K_w \tag{3.3}$$

which follows from Eq. 1.1.

If only the average specific gravity of a wide cut is known, it becomes necessary to estimate the specific gravity of the narrow fractions making up the cut. This is possible because K_w remains essentially constant for all fractions of a cut. For example, consider a petroleum crude that has a $K_w = 11.5$. It has been found that all its constituent fractions will also have $K_w \approx 11.5$, and thus Eq. 3.3 can be used to estimate S for the fractions. (The near constancy of K_w is assumed to apply to coal liquids as well, although it has not yet been proven.)

Sometimes, the specific gravity of the individual fractions is known and

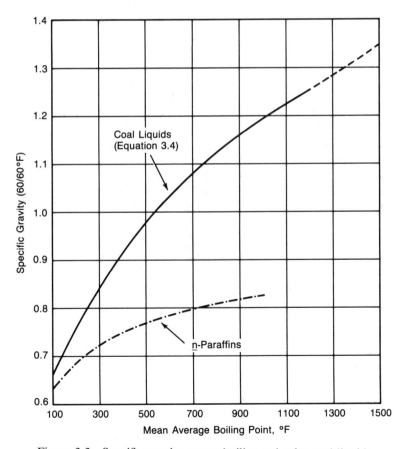

Figure 3.3   Specific gravity versus boiling point for coal liquids.

we want to estimate the specific gravity of the wide cut. In that case, it is assumed that the fractions mix ideally and the volume (the inverse of the density) of the cut is equal to the mole average volume of the fractions.

Although we stated that at least one specific gravity should be known, even that minimal information can be dispensed with. If the boiling point is known and the fraction is a coal liquid (especially, EDS or SRC-II coal liquids), then Eq. 3.4 gives a reasonable estimate of S:

$$S = 0.553461 + 1.15156 \, (t_b/1000)$$
$$- 0.708142 \, (t_b/1000)^2$$
$$+ 0.196237 \, (t_b/1000)^3 \qquad (3.4)$$

Equation 3.4 has been plotted in Figure 3.3, where the curve has been extrapolated to $t_b = 1500°F$ and $S = 1.35$. The reasonableness of Eq. 3.4, which was obtained by analyzing data on EDS and SRC-II coal liquids, is demonstrated in Table 3.2 and by the parity plot in Figure 3.4. To

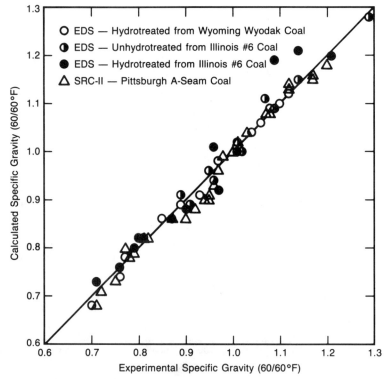

Figure 3.4   Specific gravity of coal liquids with Eq. 3.4.

Table 3.2 Specific Gravity of Coal Liquids: Correlation Deviations[a]

Correlation	EDS Coal Liquids (146 points)[b]			SRC-II Coal Liquids (90 points)[c]			Total (236 points)		
	Ave	Bias	Max	Ave	Bias	Max	Ave	Bias	Max
Equation 3.4	2.35	+0.72	+9.91	1.88	−1.06	−5.97	2.17	+0.04	+9.91
Gray and Holder (1982); Holder and Gray (1983)									
3-region model	2.58	+1.74	+12.12	0.51	−0.05	+2.30	1.80	+1.06	+12.12
2-region model	2.55	+1.68	+11.77	0.52	−0.05	+2.30	1.77	+1.02	+11.77

[a] %Dev = $100 \times (\text{calc} - \text{exp})/\text{exp}$; Ave = $[\sum_{i=1}^{N} |(\text{Dev})_i|]/N$; Bias = $[\sum_{i=1}^{N} (\text{Dev})_i]/N$; N = number of points.

[b] Hydrotreated and unhydrotreated liquids from Illinois No. 6 coal; hydrotreated liquids from Wyoming Wyodak coal. The liquids covered the ranges $120°F \leq t_b \leq 1290°F$ (estimated), and $0.70 \leq S \leq 1.29$.

[c] SRC-II liquids from Pittsburgh A-seam coal; $119°F \leq t_b \leq 969°F$ and $0.712 \leq S \leq 1.20$.

emphasize that Eq. 3.4 is good only for coal liquids, Figure 3.3 also includes the t_b/S relationship for n-paraffins. It should be clear that a correlation based only on the boiling point (or any other single property) has a limited applicability.

In examining the specific gravity-boiling point relationship, Gray and Holder (1982; see also Holder and Gray, 1983) noted a characteristic unique to SRC-II coal liquids: a rather large inflection region occurs over the temperature range 400 to 450°F, where the specific gravity is nearly constant ($S \approx 0.965$). Gray attributes this inflection region to the high concentration of phenolic compounds in this boiling range. In view of this inflection region, Gray developed for SRC-II liquids a specific model with two or three regions. Equation 3.4 does not contain this feature. However, as shown in Table 3.2, which summarizes the data base and the results of analyzing S data for EDS and SRC-II liquids, the much simpler Eq. 3.4 does a good job of fitting all the data. Indeed, Table 3.2 and Figure 3.4 suggest that there is little difference between SRC-II and EDS coal liquids, hydrotreated and unhydrotreated coal liquids, or between the various coals used. Very limited data on lignite liquids show no marked deviations from the data included in Figure 3.4.

Prediction of Molecular Weight

The molecular weight of a defined compound is its most accurately known property, but the molecular weight of a fraction must be predicted from its t_b and S. Reasonably reliable predictions (within 3 to 4%) can be made for coal liquid fractions, sometimes even when only t_b is known.

Most of the data we have used are on relatively narrow fractions ($\Delta t_b \lesssim 50$°F), and therefore all average boiling points are about the same. However, to predict M (molecular weight) of a wide cut—or to include it in a data regression—it will be necessary to convert from VABP (volume) or WABP (weight average boiling point) to MeABP. Graphs for such conversion are presented by Maxwell (1950) and the API Technical Data Book (1983).

We examined several correlations for the prediction of M. Riazi's correlation (Riazi and Daubert, 1980)

$$M = 4.5673 \times 10^{-5}(T_b)^{2.1962}(S)^{-1.0164} \tag{3.5}$$

is plotted in Figure 3.5. Although it was based only on data for defined hydrocarbons, it works very well for Illinois, Wyoming, and Pittsburgh coal liquids, as shown in Figure 3.6, a parity plot that also represents the agreement for the limited data on lignite liquids. (If the molecular weight of

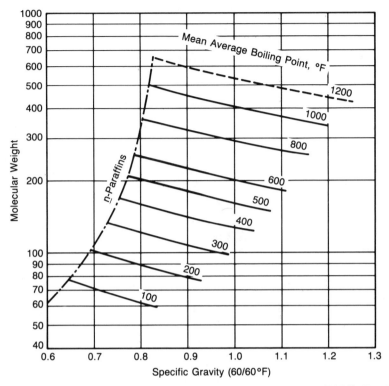

Figure 3.5   Molecular weight correlation of Riazi and Daubert (1980), Eq. 3.5.

a wide cut is known, then the T_b that fits the known M would be the MeABP of the wide cut.)

Gray (1981), Gray et al. (1983), and Gray and Holder (1982) slightly modified Eq. 3.5 to obtain a better fit of their SRC-II data:

$$M = 4.3825 \times 10^{-5}(T_b)^{2.2022}(S)^{-0.9800} \tag{3.6}$$

Another correlation for M we examined was developed by Starling's group, who modified the Kesler and Lee (1976) expression for petroleum fractions to make it applicable to coal liquids (Brulé et al., 1982):

$$
\begin{aligned}
M = {} & -12421.7 + 9316.25S + (7.753212 - 5.362614S)T_b \\
& + (1.0 - 0.753344S - 0.0173543S^2)(1.42072 \\
& - 405.3994/T_b)(5.5556 \times 10^6/T_b) + (1.0 - 0.88972S \\
& + 0.118591S^2)(1.66192 - 46.75250/T_b)(1.714678 \times 10^{11}/T_b^3)
\end{aligned}
\tag{3.7}
$$

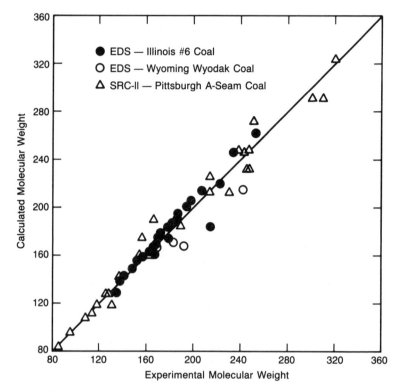

Figure 3.6   Molecular weight of coal liquids with Eq. 3.5.

In Eq. 3.7, T_b is in kelvins. (In Eq. 3.7, S may also be the specific gravity at 68/68°F, which is about 0.3% different from that at 60/60°F.)

Finally for coal liquids of unknown S, M can be predicted with the following equation:

$$\ln M = 3.91434 + 3.32452(t_b/1000)$$
$$- 2.17723(t_b/1000)^2$$
$$+ 0.776121(t_b/1000)^3 \tag{3.8}$$

Equation 3.8, which has been plotted in Figure 3.7, is based on molecular weight data for EDS and SRC-II coal liquids. The data base covered coal liquids with boiling points from 152 to 969°F and molecular weights from 85 to 363, but the curve in Figure 3.7 has been extrapolated to a molecular weight of 750 at 1500°F. Figure 3.7 also includes the t_b/M relationship for n-paraffins to emphasize that Eq. 3.8 applies only to coal liquids.

Table 3.3 compares Eqs. 3.5, 3.6, 3.7, and 3.8 against EDS and SRC-II data. Riazi's correlation, Eq. 3.5, is the best, while Eq. 3.7 is the worst.

Table 3.3 Molecular Weight of Coal Liquids: Correlation Deviations[a]

Correlation	EDS Coal Liquids (146 points)[b]			SRC-II Coal Liquids (90 points)[c]			Total (236 points)		
	Ave	Bias	Max	Ave	Bias	Max	Ave	Bias	Max
Equation 3.5	3.32	−0.21	−13.90	4.17	+0.09	+14.59	3.71	−0.07	+14.59
Equation 3.6	3.32	−0.28	−13.70	4.20	+0.09	+14.61	3.73	−0.11	+14.61
Equation 3.7	4.03	−3.87	−16.92	4.52	−1.60	−12.89	4.25	−2.83	−16.92
Equation 3.8	2.96	+0.49	−9.74	5.48	+1.26	+14.67	4.11	+0.84	+14.67

[a]See Table 3.2.

[b]Hydrotreated and unhydrotreated liquids from Illinois No. 6 coal; hydrotreated liquids from Wyoming Wyodak coal. The liquids covered the ranges $375 \leq t_b \leq 770°F$, $0.924 \leq S \leq 1.07$, and $135 \leq M \leq 253$.

[c]SRC-II liquids from Pittsburgh A-seam coal; $152 \leq t_b \leq 969°F$, $0.723 \leq S \leq 1.20$, and $85 \leq M \leq 363$.

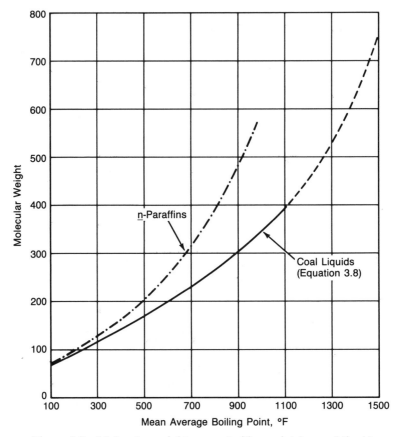

Figure 3.7   Molecular weight versus boiling point for coal liquids.

There is more on the characterization of fractions in Chapter 10. The following section concerns model compounds and the prediction of their critical constants.

MODEL COMPOUNDS

The thermodynamic properties of pure chemical substances, and therefore also of the model compounds, can best be correlated by various corresponding-states methods. Such methods require knowledge of three properties, at least for nonpolar or weakly polar compounds: the critical temperature, t_c, the critical pressure, P_c, and the acentric factor, ω.

Rigorous calculation of the acentric factor involves the vapor pressure, which is discussed in Chapter 4.

Several methods are available for predicting t_c and P_c of defined compounds; for example, see Reid et al. (1977). Here we are primarily interested in predicting t_c and P_c of coal liquid fractions, so that we can use corresponding-states methods to predict the properties of coal liquids. The development of such t_c and P_c correlations must be based on data for model compounds; that is, defined compounds characteristic of coal liquids.

As noted, although fractions have a boiling point range, their behavior should be similar to that of pure compounds when this range is narrow (say, $\Delta t_b \lesssim 50°F$). Thus, a correlation developed for pure compounds can also be expected to apply to narrow-boiling fractions. To develop t_c and P_c correlations, we use the pure compounds and data given in Table 3.4 (Wilson et al., 1981). The compounds listed in Table 1.1 are more appropriate model compounds, but the critical constants of most of them are not known—and may not be measurable because of their high critical temperatures, where the compounds could be thermally unstable.

Critical Temperature

Nokay (1959) proposed a correlation for critical temperatures that involves only t_b and S. The algebraic form of Nokay's equation was used to regress the t_c data in Table 3.4 with the following results:

$$T_c = 14.3516(T_b)^{0.66709}(S)^{0.38882} \tag{3.9}$$

For all 28 compounds, as shown in Table 3.5, Eq. 3.9 gives an average deviation of 0.66% (0.80% for the two-ring compounds) and a bias of -0.01% ($+0.13\%$ for the two-ring compounds).

Although Eq. 3.9 fits the t_c data for aromatic hydrocarbons very well, it has a limited range of applicability. For this reason, the data base was expanded by including the t_c of the C_5-C_{20} n-paraffins (American Petroleum Institute, 1983). The resulting equation

$$\log_{10} T_c = 1.29728 + 0.61954 \log_{10} T_b$$
$$+ 0.48262 \log_{10} S + 0.67365(\log_{10} S)^2 \tag{3.10}$$

is plotted in Figure 3.8, where the lines for $t_b \geq 600°F$ are extrapolations.

Table 3.5 also lists the results obtained with the equation of Riazi and Daubert (1980), which is recommended in the API Technical Data Book (1983) for petroleum fractions,

$$T_c = 24.2787(T_b)^{0.58848}(S)^{0.3596} \tag{3.11}$$

Table 3.4 Data Base for Correlation of Critical Constants of Aromatics

Compound	Normal Boiling Point (°F)[a]	Specific Gravity at 60/60°F[a]	Watson Characterization Factor K_w[a,b]	Critical Temperature (°F)[a]	Critical Pressure (psia)[a]
Benzene	176.2	0.8844	9.73	552.2	710.4
Toluene	231.1	0.8718	10.15	605.6	595.9
Ethylbenzene	277.2	0.8718	10.37	651.2	523.5
o-Xylene	292.0	0.8848	10.28	675.0	541.4
m-Xylene	282.4	0.8687	10.43	651.0	513.6
p-Xylene	281.1	0.8657	10.46	649.6	509.2
n-Propylbenzene	318.6	0.8665	10.62	689.4	464.1
Isopropylbenzene	306.3	0.8663	10.57	676.4	465.4
1-Methyl-2-ethylbenzene	329.3	0.8851	10.45	712	440.9
1-Methyl-3-ethylbenzene	322.4	0.8690	10.61	687	411.5
1-Methyl-4-ethylbenzene	323.6	0.8657	10.65	693	426.2
1,2,3-Trimethylbenzene	349.0	0.8987	10.37	736.5	501.0
1,2,4-Trimethylbenzene	336.9	0.8803	10.54	708.8	468.8
1,3,5-Trimethylbenzene	328.5	0.8696	10.63	687.6	453.5
n-Butylbenzene	362.0	0.8646	10.84	729.3	418.7
Isobutylbenzene	343.0	0.8576	10.84	711	440
sec-Butylbenzene	344.0	0.8664	10.74	736.5	428.0

34

tert-Butylbenzene	336.5	0.8710	10.65	728.0	430.0
1-Methyl-2-isopropylbenzene	352.7	0.8810	10.60	746.0	420.0
1-Methyl-3-isopropylbenzene	347.1	0.8654	10.76	739.5	426.2
1-Methyl-4-isopropylbenzene	350.8	0.8617	10.83	716.0	410
Styrene	293.3	0.9110	9.99	706.0	580.0
Naphthalene	424.4	1.030[c]	9.32	887.5	587.5
1-Methylnaphthalene	472.4	1.0244	9.536	930.0	503.9[d]
Indan (2,3-dihydroindene)	352.1	0.9685	9.63	773.1[e]	573[e]
Tetralin	405.7	0.9739	9.785	830[f]	524.4[d]
Phenylbenzene	491.0	1.027[c]	9.57	961[g]	558[g]
Diphenylmethane	507.7	1.0104[c]	9.79	926[h]	415[h]

[a] All values from API's *Technical Data Book—Petroleum Refining*, 4th ed. (1983), unless otherwise indicated. Table 3.4 is identical to that reported by Wilson et al. (1981), except where inconsistencies with Table 1.1 were corrected (for naphthalene, 1-methylnaphthalene, tetralin, and phenylbenzene).

[b] $K_w = T_b(°R)^{1/3}/S$.

[c] For supercooled liquid below the normal freezing point (extrapolated).

[d] Estimated by extrapolating experimental vapor pressure data to the critical temperature.

[e] Ambrose et al. (1974).

[f] Estimated by Lydersen's (1955) procedure.

[g] Kudchadker et al. (1968).

[h] Guye and Mallet (1902).

Table 3.5 Prediction of Critical Temperature and Critical Pressure of Aromatics

Correlation	Ave	Bias	Max
	Deviations in Critical Temperature[a]		
Equation 3.9	0.66	−0.01	+2.01
Equation 3.10	0.70	−0.02	−1.92
Equation 3.11	1.11	+0.35	+2.48
Equation 3.12	0.98	−0.39	−3.11
Equation 3.13	1.01	−0.40	−3.39
	Deviations in Critical Pressure[a]		
Equation 3.14	$\{$(4.54	+0.25	+9.74)[b]
	(3.63	+0.17	+13.83)[c]
Equation 3.15	3.47	+0.25	+10.89
Equation 3.16	5.23	−1.52	−22.20

[a]See Table 3.2.

[b]With experimental t_c.

[c]With calculated t_c (Eq. 3.9).

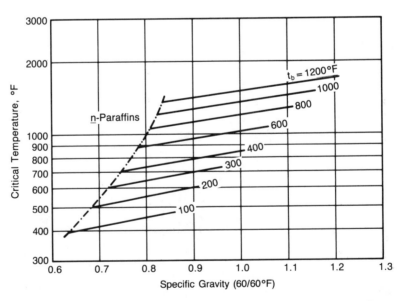

Figure 3.8 Critical temperature calculated from boiling point and specific gravity (Eq. 3.10).

and the equation by Starling's group (Brulé et al., 1982):

$$
\begin{aligned}
T_c = \ & 429.138 + 0.886861\, t_b - 4.596433 \times 10^{-4} t_b^2 \\
& - 2.410089 \times 10^{-3}\text{API} \times t_b + 1.630489 \times 10^{-7} t_b^3 \\
& - 9.323778 \times 10^{-7}\text{API} \times t_b^2 - 1.430628 \times 10^{-8}\text{API}^2 \times t_b^2
\end{aligned} \quad (3.12)
$$

In Eq. 3.12, T_c is in kelvins, but t_b is in degrees Fahrenheit. (Equation 3.2 converts °API to S.) We have also included in the comparison the equation recommended in the API Technical Data Book (1983) for aromatic compounds:

$$
T_c = 13.8497(T_b)^{0.66929}(S)^{0.22732} \quad (3.13)
$$

(This differs from Eq. 3.9 mainly in the exponent of the S term.)

Though Eq. 3.10 is slightly inferior to Eq. 3.9 for aromatics, it was selected for predicting critical temperatures of narrow-boiling aromatic fractions because it has a wider range of applicability. When the C_5–C_{20} n-paraffins are included in the comparison, the average deviation is 0.54% with Eq. 3.10 and 0.94% with Eq. 3.9.

Critical Pressure

Wilson et al. (1981) described the development of a correlation for the P_c of aromatic fractions, again based on the data in Table 3.4. The resulting correlation expresses P_c (in psia) as a function of T_b, S, and T_c:

$$
\log_{10} P_c = 2.22066 - 0.05445 K_w + 3.12579\left(1 - \frac{T_b}{T_c}\right) \quad (3.14)
$$

Equation 3.10 (or 3.9) should be used to predict T_c if it is not known. A correlation of wider applicability was obtained by including the P_c data for the C_5–C_{20} n-paraffins (American Petroleum Institute, 1983):

$$
\begin{aligned}
\log_{10} P_c = \ & 9.08740 - 2.15833 \log_{10} T_b + 3.35417 \log_{10} S \\
& + 5.64019(\log_{10} S)^2
\end{aligned} \quad (3.15)
$$

Equation 3.15 is plotted in Figure 3.9. As in the case of Figure 3.8, the lines for $t_b \geq 600°$F are extrapolated. An even better agreement with the data can be obtained by adding a $(\log_{10} T_b)^2$ term to Eq. 3.15 (also true for Eq. 3.10), but the limited data did not justify such an addition.

Table 3.5 lists the results of the analysis of the P_c data in Table 3.4 with Eqs. 3.14, 3.15, and the relationship of Riazi and Daubert (1980), which is also given in the API Technical Data Book (1983):

$$
P_c = 3.12281 \times 10^9 (T_b)^{-2.3125}(S)^{2.3201} \quad (3.16)
$$

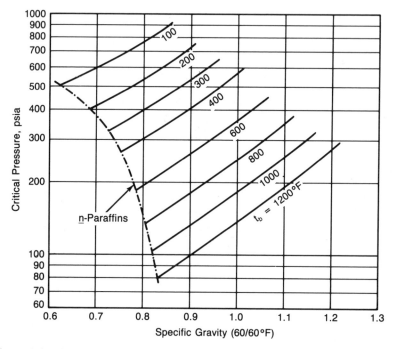

Figure 3.9   Critical pressure from boiling point and specific gravity (Eq. 3.15).

The data analysis shows that Eq. 3.15 is the best, notwithstanding its noted limitations, and it will therefore be used to predict the critical pressure of narrow-boiling aromatic fractions. However, a better approach is to predict P_c at t_c with a (reliable) vapor pressure correlation, and thus make P_c consistent with the vapor pressure. Vapor pressure is discussed in Chapter 4.

Nonhydrocarbon model compounds were not used in developing the t_c and P_c correlations because a high level of heteroatoms could not be taken properly into account when only t_b and S are used as correlating parameters. There is more on heteroatoms in later chapters, but it should be noted here that, although coal liquids are rich in oxygen, they still have less oxygen than some oxygen-containing model compounds. For example, phenol has 17 wt% oxygen and 1-naphthol 11 wt%, while coal liquids generally have less than 4 wt% (see Chapter 1). Thus, oxygen should have much less influence on the properties of coal liquids than it does on phenol or 1-naphthol.

Acentric Factor

The acentric factor, ω, has gained very wide acceptance as the third parameter in the extended corresponding-states principle. It is defined as

$$\omega = -\log_{10}(P_r^s)_{T_r=0.7} - 1.0 \tag{3.17}$$

where P^s is the vapor pressure and r designates a reduced property ($T_r = T/T_c$; $P_r = P/P_c$). To calculate ω, we need, in addition to T_c and P_c, the vapor pressure of the model compound or fraction. This is the subject of Chapter 4.

As an alternative to using Eq. 3.17 with the vapor pressure at $T_r = 0.7$, ω can be estimated with a relation first suggested by Edmister (1958):

$$\omega = \frac{3}{7}\left(\frac{T_{b,r}}{1 - T_{b,r}}\right)\log_{10}(P_c/14.696) - 1.0 \tag{3.18}$$

For nonpolar fluids, the values obtained with Eq. 3.18 are in good agreement with those obtained from Eq. 3.17. Unlike Eq. 3.17, Eq. 3.18 cannot be used as $T_c \rightarrow T_b$ because it becomes indeterminate ($\omega \rightarrow 0/0$).

RECOMMENDATIONS

For convenient reference, this section summarizes the recommendations made in Chapter 3 for the characterization of coal-liquid fractions and model compounds.

It is assumed that the average boiling point and specific gravity of a coal-liquid fraction are known. If the specific gravity is unavailable, it can be estimated with Eq. 3.4 or read off Figure 3.3.

For the molecular weight of a fraction, use Eq. 3.5, which is illustrated in Figure 3.5. If S is not known, Eq. 3.8 or Figure 3.7 can be used.

The critical temperature of a fraction can best be estimated with Eq. 3.10 or Figure 3.8. Equation 3.10 is preferred over Eq. 3.9 because it has a wider range of applicability.

The calculation of critical pressure can best be made with a vapor pressure correlation that is valid up to the critical temperature (see Chapter 4). As an alternative, Eq. 3.15 or Figure 3.9 can be used.

Vapor pressure is also needed for the calculation of the acentric factor with Eq. 3.17, which is the preferred method. For most nonpolar systems, however, Edmister's relation, Eq. 3.18, provides a satisfactory approximation.

With the characterization of coal liquids and model compounds completed, the following six chapters examine their properties, in approximate order of priority.

REFERENCES

Ambrose, D., B. E. Broderick, and R. Townsend, The Critical Temperatures and Pressures of Thirty Organic Compounds, *J. Appl. Chem. Biotechnol.*, **24**, 359 (1974).

American Petroleum Institute, *Technical Data Book—Petroleum Refining*, 4th Ed., API, Washington, DC, 1983.

Brulé, M. R., C. T. Lin, L. L. Lee, and K. E. Starling, Multiparameter Corresponding-States Correlation of Coal-Fluid Thermodynamic Properties, *AIChE J.*, **28**, 616 (1982).

Edmister, W. C., Compressibility Factors and Equations of State, *Petrol. Refiner*, **37**(4), 173 (1958).

Gray, J. A., Selected Physical, Chemical, and Thermodynamic Properties of Narrow Boiling Range Coal Liquids from the SRC-II Process, Report No. DOE/ET/10104-7, April 1981.

Gray, J. A., C. J. Brady, J. R. Cunningham, J. R. Freeman, and G. M. Wilson, Thermophysical Properties of Coal Liquids. 1. Selected Physical, Chemical, and Thermodynamic Properties of Narrow Boiling Range Coal Liquids, *Ind. Eng. Chem. Process Des. Dev.*, **22**, 410 (1983).

Gray, J. A., and G. D. Holder, Selected Physical, Chemical, and Thermodynamic Properties of Narrow Boiling Range Coal Liquids from the SRC-II Process, Supplementary Property Data, Report No. DOE/ET/10104-44, April 1982.

Green, L. E., Chromatograph Gives Boiling Point, *Hydrocarbon Process.*, **55**(5), 205 (1976).

Guye, P. A., and E. Mallet, Measurement of Critical Constants, *Arch. Sci. Phys. Nat.*, **13**(4), 30, 274 (1902); in French.

Holder, G. D., and J. A. Gray, Thermophysical Properties of Coal Liquids. 2. Correlating Coal Liquid Densities, *Ind. Eng. Chem. Process Des. Dev.*, **22**, 424 (1983).

Kesler, M. G., and B. I. Lee, Improve Prediction of Enthalpy Fractions, *Hydrocarbon Process.*, **55**(3), 153 (1976).

Kudchadker, A. P., G. H. Alani, and B. J. Zwolinski, The Critical Constants of Organic Substances, *Chem. Rev.*, **68**, 659 (1968).

Lydersen, A. L., Estimation of Critical Properties of Organic Compounds, *Univ. Wisconsin Eng. Exp. Sta. Rept.*, **3**, Madison (April 1955).

Maxwell, J. B., *Data Book on Hydrocarbons*, Van Nostrand, Princeton, NJ, 1950.

Nokay, R., Estimate Petrochemical Properties, *Chem. Eng.*, **66**(4), 147 (1959).

Pannell, R. B., and A. Sood, Simulated Distillation of Coal Liquids, *J. Chromat. Sci.*, **20**, 433 (1982).

Reid, R. C., J. M. Prausnitz, and T. K. Sherwood, *The Properties of Gases and Liquids*, 3rd Ed., McGraw-Hill, New York, 1977.

Riazi, M. R., and T. W. Daubert, Simplify Property Predictions, *Hydrocarbon Process.*, **59**(3), 115 (1980).

Wilson, G. M., R. H. Johnston, S. C. Hwang, and C. Tsonopoulos, Volatility of Coal Liquids at High Temperatures and Pressures, *Ind. Eng. Chem. Process Des. Dev.*, **20**, 94 (1981).

Chapter 4

VAPOR PRESSURE

The importance of vapor pressure and vapor-liquid equilibria in engineering design was emphasized in Chapter 2. These two topics are handled separately: vapor pressure in this chapter and vapor-liquid equilibria in Chapter 5. In both cases, the emphasis is on high temperatures, up to 900°F.

The vapor pressure of petroleum fractions can be reliably predicted with the Maxwell-Bonnell (1957) correlation. This correlation only requires two input parameters, t_b and K_w. On the other hand, the vapor pressure of defined compounds is predicted here with a modified form of Riedel's (1954) equation. This equation can also be used for fractions if their t_c, P_c, and ω (acentric factor) are known.

The MB correlation is examined first. We have already demonstrated (Wilson et al., 1981) that this correlation can be significantly in error at superatmospheric pressures when it is applied to polynuclear aromatics or highly aromatic liquids such as coal liquids; that is, liquids with $K_w \lesssim 10$. A modification of the MB correlation for such liquids is presented. An alternative to MB, the Lee-Kesler (1980) correlation, which uses t_b and S as its parameters, is examined next. Finally, the modified Riedel equation, which is at its best between t_b and t_c, is considered.

After the three correlations, we examine experimental data for model compounds and relatively narrow coal liquid fractions, with particular emphasis on superatmospheric conditions. One issue of interest that is resolved is the equivalence between narrow fractions and defined compounds. The chapter concludes with recommendations for calculating vapor pressures and acentric factors.

MAXWELL-BONNELL CORRELATION

The MB correlation has been used in the petroleum industry since the mid-fifties (Maxwell and Bonnell, 1955, 1957); see also the API Technical Data Book (1983). It is an industry standard for converting vacuum distillations to atmospheric distillations; that is, for converting subatmospheric boiling points to normal boiling points. Since laboratory distillations generally are not carried out above 650°F, higher normal boiling points are determined by extrapolation with the MB correlation.

The MB correlation, which uses n-hexane ($K_w = 12.8$) as the reference compound, is given graphically in Figures 4.1–4.5 (reprinted from Maxwell and Bonnell, 1955). These figures can be used without any correction when $K_w = 12.0$, a typical value for petroleum crude fractions. An algebraic form of the MB correlation, also for $K_w = 12.0$, is given by the following equation for the boiling point of a fraction when the vapor pressure of the fraction is equal to that of n-hexane:

$$T_b = \frac{748.1(A + 0.0002867)}{(1/T) - 0.0002867 + 0.2145(A + 0.0002867)} \qquad (4.1)$$

The parameter A is a function of pressure given by

$$A = \frac{5.9082 - \log_{10} P^s}{2926.8526 - 43 \log_{10} P^s}, \qquad P^s < 0.0387 \text{ psia (2 mm Hg)} \qquad (4.2a)$$

$$A = \frac{5.0442 - \log_{10} P^s}{2499.0330 - 95.76 \log_{10} P^s}, \qquad 0.0387 \le P^s \le 14.696 \text{ psia} \qquad (4.2b)$$

$$A = \frac{5.4932 - \log_{10} P^s}{2708.3952 - 36.0 \log_{10} P^s}, \qquad P^s > 14.696 \text{ psia} \qquad (4.2c)$$

A formulation of the MB correlation with P^s as the explicit variable is given in the API Technical Data Book (1983).

For hydrocarbons with K_w substantially different from 12.0, a correction is necessary to the boiling point calculated with Eqs. 4.1–4.2. This correction is given by

$$\begin{aligned} \Delta t_b &= t_{b,\text{actual}} - t_{b,\text{MB @ } K_w = 12.0} \\ &= F \times 2.5(K_w - 12.0) \times \log_{10}(P^s/14.696) \end{aligned} \qquad (4.3)$$

where

$$\begin{aligned} F &= 0 & t_b &\le 200°\text{F} & (4.4a) \\ &= -1 + 0.005 \, t_b, & 200 &\le t_b \le 400°\text{F} & (4.4b) \\ &= 1, & t_b &> 400°\text{F} & (4.4c) \end{aligned}$$

Equation 4.4 (American Petroleum Institute, 1983) gives only slightly different values for F from those in Figure 4.6 (Maxwell and Bonnell, 1957), which is a plot of Eq. 4.3 with $F = 1$.

As noted by Wilson et al. (1981), MB at subatmospheric pressures was found to be satisfactory even for coal liquids. Problems were discovered at superatmospheric pressures. It was decided to leave the correlation at

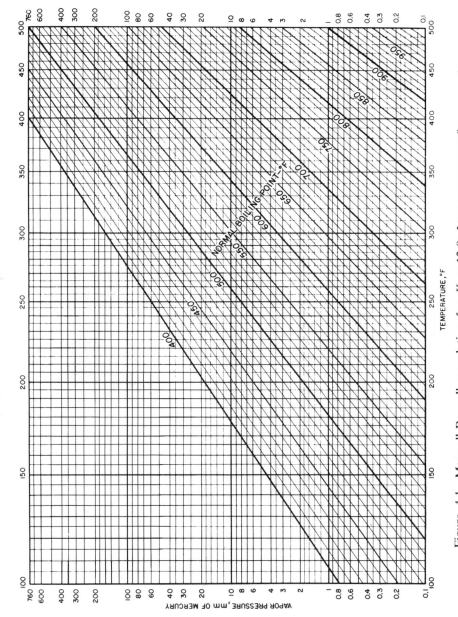

Figure 4.1 Maxwell-Bonnell correlation for $K_w = 12.0$. Low temperature/low normal boiling point range.

43

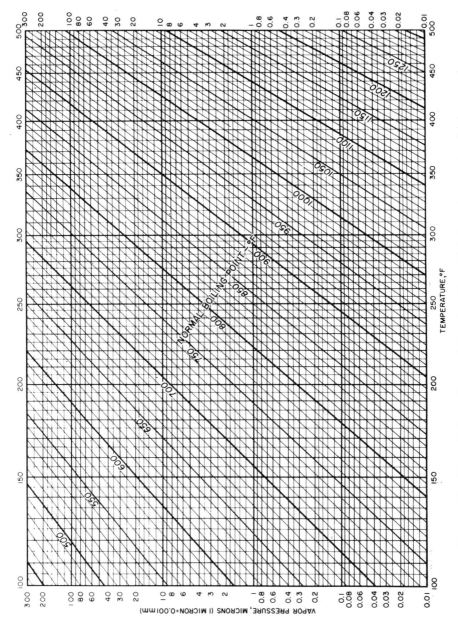

Figure 4.2 Maxwell-Bonnell correlation for $K_w = 12.0$. Low temperature/high normal boiling point range.

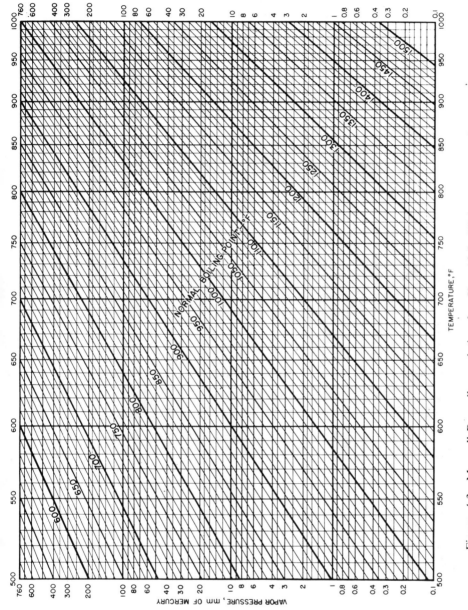

Figure 4.3   Maxwell-Bonnell correlation for $K_w = 12.0$. High temperature/low normal boiling point range.

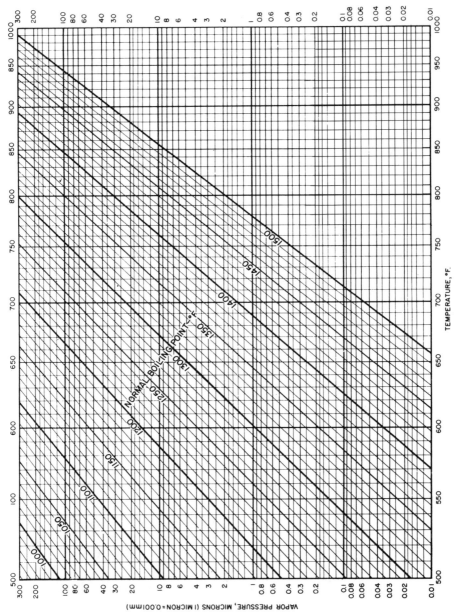

Figure 4.4 Maxwell-Bonnell correlation for $K_w = 12.0$. High temperature/high normal boiling point range.

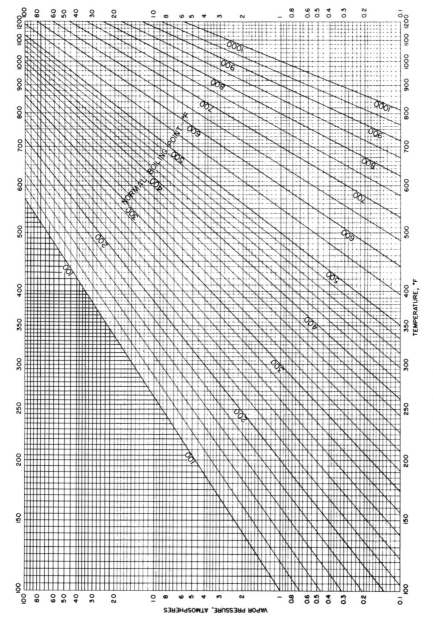

Figure 4.5   Maxwell-Bonnell correlation for $K_w = 12.0$ at superatmospheric conditions.

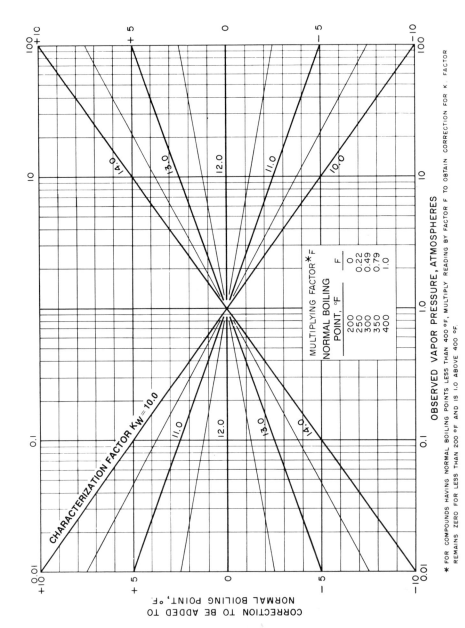

OBSERVED VAPOR PRESSURE, ATMOSPHERES

CORRECTION TO BE ADDED TO NORMAL BOILING POINT, °F.

NORMAL BOILING POINT, °F	F
200	0
250	0.22
300	0.49
350	0.79
400	1.0

MULTIPLYING FACTOR* F

CHARACTERIZATION FACTOR Kw = 10.0

* FOR COMPOUNDS HAVING NORMAL BOILING POINTS LESS THAN 400 °F, MULTIPLY READING BY FACTOR F TO OBTAIN CORRECTION FOR K. FACTOR REMAINS ZERO FOR LESS THAN 200 °F AND IS 1.0 ABOVE 400 °F.

Figure 4.6 Boiling point correction for $K_w \neq 12.0$.

$K_w = 12.0$ untouched, and instead to focus on the correction for the range $K_w < 12.0$, which covers all the coal liquids of interest.

Problems with the t_b Correction for $K_w < 12.0$

Figure 4.7 is a specific case of Figure 4.6 for tetralin ($t_b = 405.7°F$, $K_w = 9.78$). Data from various sources demonstrate that, with some exceptions, MB works well up to about 3 atm (44 psia), but then it progressively gets worse. The linear dependence of Δt_b on log P^s is reasonable for $P^s < 3$ atm, but clearly it is unacceptable at higher pressures.

A more surprising comparison is shown in Figure 4.8 for anthracene ($t_b = 647.7°F$, $K_w = 9.21$). In this case, MB is unsatisfactory even at sub-atmospheric pressures. However, Δt_b remains linear below 1 atm. In view of this, a simple correction was to use Eq. 4.4b to determine F for $t_b > 400°F$, rather than set $F = 1.0$. As shown by the dotted line in Figure 4.8, the improvement at subatmospheric pressure was considerable. Since that was not sufficient at superatmospheric pressures, however, a modification of the correlation was developed.

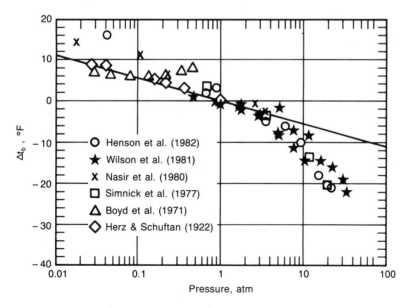

Figure 4.7   Boiling point correction for tetralin ($t_b = 405.7°F$; $K_w = 9.78$) according to Eqs. 4.3 and 4.4.

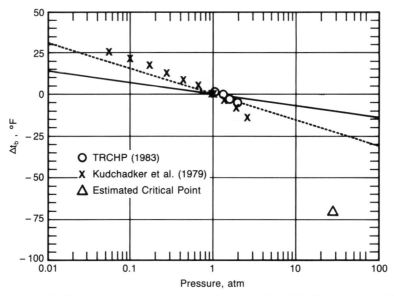

Figure 4.8   Boiling point correction for anthracene ($t_b = 647.7°F$; $K_w = 9.21$) according to Eqs. 4.3 and 4.4 (———) or Eqs. 4.3 and 4.4b (----).

Modification of the t_b Correction for $K_w < 12.0$

The final form of the t_b correction selected was

$$\Delta t_b = f_1(t_b) \times f_2(K_w) \times f_3(P) \tag{4.5}$$

A total of 957 points for model compounds (including alkylbenzenes) and coal-liquid fractions, 488 subatmospheric and 469 superatmospheric, covering the range $8.3 \le K_w \le 11.1$ were used in the data regression. *No heteroatom-containing compounds were included in the data base.*

First of all, Eq. 4.4 was used for $f_1(t_b)$ or F, even above 400°F:

$$\begin{aligned} f_1(t_b) &= 0, & t_b &\le 200°F \\ &= -1 + 0.005\, t_b, & t_b &> 200°F \end{aligned} \tag{4.6a}$$

The data analysis demonstrated the need for a quadratic term in the K_w dependence. Accordingly, $f_2(K_w)$ is given by

$$f_2(K_w) = (K_w - 12.0) - 0.01304(K_w - 12.0)^2 \tag{4.6b}$$

Finally, Figure 4.7 clearly shows that the pressure dependence in the superatmospheric range also requires a quadratic term. The final form of

Table 4.1 Deviations in Calculated Boiling Points of Aromatic Compounds and Fractions ($K_w < 12.0$)

Method	$P^s \leq 1$ atm (488 points)[a]			$P^s > 1$ atm (469 points)[a]			Overall (957 points)[a]		
	Ave	Bias	Max	Ave	Bias	Max	Ave	Bias	Max
Original MB	3.00	−1.86	−36.37	5.86	+5.51	+60.74	4.40	+1.75	+60.74
MB with new $f_1(t_b)$—Eq. 4.6a	2.39	−0.32	−21.45	5.08	+4.54	+48.24	3.71	+2.06	+48.24
MB with new Δt_b—Eqs. 4.5–4.6	2.47	+0.11	−19.91	3.54	+0.67	+27.73	2.99	+0.38	+27.73

[a]Dev $= \Delta t_{b,exp} - \Delta t_{b,cal}$, in °F; Ave $= \sum |\text{Dev}|/N$; Bias $= \sum \text{Dev}/N$; $N =$ number of points.

$f_3(P)$ is

$$f_3(P) = 2.6536 \log_{10}P, \qquad\qquad\qquad P \le 1 \text{ atm}$$
$$= 2.6536 \log_{10}P + 2.1435 (\log_{10}P)^2, \quad P > 1 \text{ atm} \qquad (4.6c)$$

In Eq. 4.6c, P is in atmospheres.

The results of the data analysis are presented in Table 4.1. The modified Δt_b correlation given by Eqs. 4.5–4.6 or Figure 4.9 improved the fit of the subatmospheric data by about 18%, of the superatmospheric data by about 40%, and of all the data by about 32%. This is substantial improvement, especially considering that the MB correlation for $K_w = 12.0$, Eqs. 4.1–4.2, was left untouched. The use of an aromatic reference compound, such as benzene ($K_w = 9.7$) or naphthalene ($K_w = 9.3$), in place of n-hexane ($K_w = 12.8$), should definitely lead to an even better vapor pressure correlation for aromatic compounds and fractions. Still, the modified MB correlation for $K_w < 12$ presented here is considered to be of acceptable accuracy for engineering work. It is compared with other correlations later in this chapter.

Although the MB correlation for $K_w = 12$ was not altered, there is

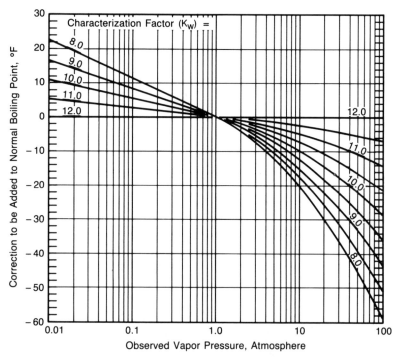

Figure 4.9   Modified boiling point correction for $K_w < 12$.

evidence that a correction is needed for $K_w = 12$ (as well as for $K_w > 12$). When superatmospheric data are plotted versus temperature, it becomes clear that the focal point, the point where all the vapor pressure lines converge, should be at a much higher temperature and pressure than that in the MB correlation. Such a higher focal point was proposed by Lee and Kesler (1980). A similar conclusion was also reached by Zudkevitch et al. (1983).

LEE-KESLER CORRELATION

Lee and Kesler (1980) presented an alternative to MB that also uses n-hexane as its reference compound. For n-paraffins and most branched paraffins, Lee and Kesler proposed the following relationship:

$$\ln P_n^s = \left(0.1047 + \frac{259.8}{T_e + 55}\right) T_e - 6.074(T_e)^{0.5} - 140.65 \tag{4.7}$$

In the Lee-Kesler correlation, T is in kelvins and P in mm Hg. In Eq. 4.7, P_n^s is the vapor pressure of n-paraffins and T_e is the temperature for n-hexane or the temperature at which other paraffins have the same vapor pressure as n-hexane, as given by

$$T_e = 341.9\left[\frac{(T_b + C)^{-1} - B}{(T + C)^{-1} - B}\right] \tag{4.8a}$$

where

$$B = 1.09691\left(\frac{1}{3870 + C} - \frac{0.088346}{T_b + C}\right) \tag{4.8b}$$

and

$$C = -10.15 + 0.15819(T_b) - 3.75841 \times 10^{-4}(T_b)^2 \tag{4.8c}$$

For non-paraffins, the deviation from Eq. 4.7 can be substantial. This departure is represented as follows:

$$\ln P^s = \ln P_n^s + \left[\left(\frac{S}{S_n}\right)^{0.5} - 1\right]\left[\left(\frac{T_b}{T}\right)^3 - 1\right]\left[3.214 - 3.765\left(\frac{T_b'}{T_b}\right)^2\right] \tag{4.9}$$

T_b' is the normal boiling point of n-hexane, 341.9 K, and S_n, the specific gravity of n-paraffins at 60/60°F, is given by

$$S_n = 1.0475 - 1.511 \times 10^{-4}(T_b) + 7.127 \times 10^{-8}(T_b)^2 - 116.4/T_b \tag{4.10}$$

Lee and Kesler evaluated their correlation, Eqs. 4.7–4.10, against a very large data base for C_2–C_{100} hydrocarbons. They found their correlation to be a significant improvement over MB, especially for C_{31}–C_{100} *n*-paraffins.

An evaluation of the Lee-Kesler (1980) correlation against model compounds and coal liquids is presented later in this chapter.

THE RIEDEL EQUATION

Wilson et al. (1981) concluded that the MB correlation was generally adequate for aromatic fractions up to t_b, but a better correlation was needed between t_b and t_c, the critical temperature. In this range, probably the best vapor pressure correlation is that proposed by Riedel (1954). Accordingly, the approach used here involves first the prediction of the critical point, as recommended in Chapter 3, and then the interpolation between that and the boiling point with a modified form of the Riedel equation.

The following modified form of Riedel's equation has been developed specifically for the range t_b to t_c:

$$\ln P_r^s = A - B/T_r - C \ln T_r + DT_r^6 \qquad (4.11)$$

where

$$A = 5.671485 + 12.439604\omega \qquad (4.12a)$$

$$B = 5.809839 + 12.755971\omega \qquad (4.12b)$$

$$C = 0.867513 + 9.654169\omega \qquad (4.12c)$$

$$D = 0.1383536 + 0.316367\omega \qquad (4.12d)$$

and ω is Pitzer's acentric factor, defined as

$$\omega = -\log_{10}(P_r^s)_{T_r=0.7} - 1.0 \qquad (4.13)$$

The subscript r denotes a reduced property ($T_r = T/T_c$; $P_r = P/P_c$).

Equation 4.11 can also be used in a nonreduced form; that is, with P^s and T in place of P_r^s and T_r, respectively, and with A–D as adjustable parameters:

$$\ln P^s = A - B/T - C \ln T + DT^6 \qquad (4.14)$$

Equation 4.14 gives an excellent fit of experimental data, but clearly is not a predictive method.

If Eq. 4.11 is used at the normal boiling point, it reduces to

$$-\ln(P_c/14.696) = A - B/T_{b,r} - C \ln T_{b,r} + DT_{b,r}^6$$

and substitution for A, B, C, D results in the following expression for ω:

$$\omega = \frac{-\ln(P_c/14.696) - 5.671485 + 5.809839/T_{b,r} + 0.867513 \ln T_{b,r} - 0.1383536 T_{b,r}^6}{12.439604 - 12.755971/T_{b,r} - 9.654169 \ln T_{b,r} + 0.316367 T_{b,r}^6} \quad (4.15)$$

T_c and P_c must be known (or be estimated) to use the Riedel equation. On the other hand, if the MB or Lee-Kesler (1980) correlations are used only T_c must be known, since the vapor pressure correlation can be used to estimate $P_c(= P^s$ at $T_c)$. This is the preferred procedure for calculating P_c and ω, so long as the MB (or the Lee-Kesler) correlation is reliable in the range $T_r = 0.7$ to 1.0.

The MB, modified MB, Lee-Kesler, and Riedel correlations are evaluated against experimental data in the following two sections.

EXPERIMENTAL DATA ON MODEL COMPOUNDS

As already discussed, the major deficiency of the MB correlation for aromatic liquids was found to be at superatmospheric pressures. Unfortunately, very little information is available at such pressures on aromatic fractions or even pure polynuclear aromatic compounds. Therefore, it became necessary to carry out an extensive experimental program on the vapor pressure of model compounds and coal liquid fractions up to 900°F (Wilson et al., 1981). That work has been complemented by the extensive investigations of K. C. Chao, at Purdue University, and Riki Kobayashi, at Rice University.

Two examples illustrate the prediction of high-temperature vapor pressures for model compounds. Figure 4.10a shows the deviations of the four generalized correlations from data on tetralin, model compound for the EDS hydrogenated solvent cut, fitted with the nongeneralized Riedel equation, Eq. 4.14. A similar comparison is shown in Figure 4.10b for 1-methylnaphthalene, model compound for the unhydrogenated solvent cut.

Wilson et al. (1981) analyzed superatmospheric vapor pressure data for 11 polynuclear aromatics and found that the modified Riedel equation was much better than the MB correlation. While the latter gave an average deviation of 6.3%, the new procedure reduced the deviation to less than 3.0%. A more extensive comparison is given in Table 4.2, where subatmospheric data are also included (down to a reduced temperature of 0.45).

Table 4.2 clearly demonstrates that, when all compounds are considered, the modified Riedel equation is the best generalized method. Surprisingly, the modified MB is slightly inferior to the original one. However, only hydrocarbons were considered in developing the modified MB. Indeed, for

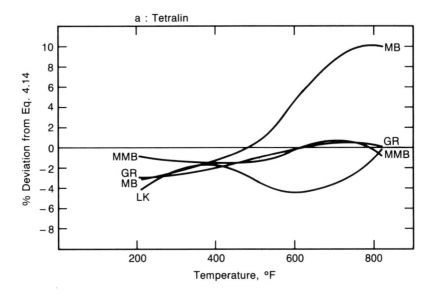

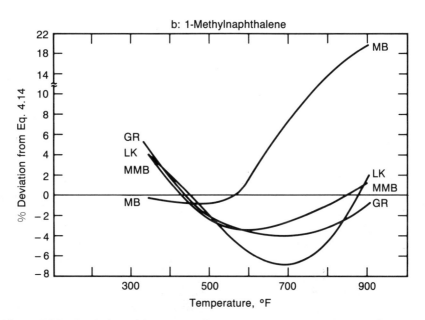

Figure 4.10 Deviation of four generalized vapor pressure correlations [MB: Maxwell-Bonnell; MMB: modified Maxwell-Bonnell; LK: Lee-Kesler (1980); GR: generalized Riedel] from non-generalized Riedel, Eq. 4.14. (a), Tetralin; (b), 1-Methylnaphthalene.

Table 4.2 Vapor Pressure Predictions for Model Compounds

| | Percent Average Deviation | | | | |
Compound (No. Points)	Maxwell-Bonnell	Modified Maxwell-Bonnell	Lee-Kesler	Modified Riedel	Data Sources
Naphthalene (70)	2.97	2.30	2.45	2.18	Crafts (1915); Ward and Van Winkle (1954); Camin and Rossini (1955); Kobayashi (1978); Wilson et al. (1981)
Tetralin (55)	4.82	2.48	3.04	2.50	Herz and Schuftan (1922); Boyd et al. (1971); Simnick et al. (1977); Nasir et al. (1980); Wilson et al. (1981); Kara et al. (1981); Henson et al. (1982)
cis-Decalin (39)	8.52	7.25	6.33	4.60	Camin and Rossini (1955); Lenoir et al. (1971); Wilson et al. (1981)
1-Methylnaphthalene (104)	3.54	3.57	3.67	3.84	Myers and Fenske (1955); Camin and Rossini (1955); Glaser and Rüland (1957); Yao et al. (1977/1978); Wieczorek and Kobayashi (1981); Wilson et al. (1981); Henson et al. (1982)
2-Methylnaphthalene[a] (29)	2.89	1.09	1.66	0.70	Glaser and Rüland (1957); Wieczorek and Kobayashi (1981)

Table 4.2 (*Continued*)

Compound (No. Points)	Percent Average Deviation				Data Sources
	Maxwell-Bonnell	Modified Maxwell-Bonnell	Lee-Kesler	Modified Riedel	
Phenylbenzene (116)	4.54	3.91	4.22	3.28	Jacquerod and Wassmer (1904); Garrick (1927); Chipman and Peltier (1929); Moor and Kanep (1936); Glaser and Rüland (1957); Nasir et al. (1980)
Phenylcyclohexane (7)	21.78	13.97	9.23	7.34	Wilson et al. (1981)
Cyclohexylcyclohexane (46)	8.23	6.41	6.94	2.36	Myers and Fenske (1955); Wieczorek and Kobayashi (1980)
Diphenylmethane (69)	3.77	3.45	4.66	1.74	Crafts (1915); Glaser and Rüland (1957); Simnick et al. (1978); Wieczorek and Kobayashi (1980); Wilson et al. (1981)
Fluorene[a] (20)	5.42	6.47	4.06	2.30	Sivaraman and Kobayashi (1982)
Phenanthrene (57)	10.88	5.98	6.44	10.51	Mortimer and Murphy (1923); Nelson and Senseman (1922); Wilson et al. (1981)
9,10-Dihydrophenanthrene (17)	5.89	10.35	5.68	7.40	Wieczorek and Kobayashi (1981)
Anthracene (17)	6.47	1.60	2.20	7.44	Kudchadker et al. (1979); TRCHP (1983)
m-Cresol[a] (47)	20.81	24.38	25.33	6.35	Stull (1947); Glaser and Rüland (1957); Simnick et al. (1979); Nasir et al. (1980)
2,4-Xylenol[a] (38)	6.61	7.92	8.40	3.15	Andon et al. (1960); Wilson et al. (1981)

9-Hydroxyfluorene[a,b] (24)	69.42	48.21	60.13	62.78	Sivaraman et al. (1983)
9-Fluorenone[a] (23)	15.92	10.01	4.77	8.35	Sivaraman et al. (1983)
Dibenzofuran[a] (19)	13.31	6.95	7.61	11.91	Sivaraman and Kobayashi (1982)
Quinoline[a] (63)	2.18	5.50	6.40	2.64	Glaser and Rüland (1957); Malanowski (1961); Kobayashi (1978); Sebastian et al. (1978a); Wilson et al. (1981)
Carbazole[a] (26)	2.91	3.72	2.95	0.80	Sivaraman et al. (1983)
Benzothiophene[a] (32)	1.53	4.60	6.02	1.40	Sebastian et al. (1978b); Wieczorek and Kobayashi (1980)
Dibenzothiophene[a] (19)	3.27	35.01	18.35	11.20	Sivaraman and Kobayashi (1982)
All compounds (913[c])	6.23	6.36	6.11	4.08	
All hydrocarbons (646)	5.44	4.24	4.32	3.79	

[a]Not included in data base used in the modification of the MB correlation.
[b]Vapor pressure measured only up to 1.6 psia (449.4°F); t_b and S in Table 1.1 are rough estimates.
[c]Excludes 9-hydroxyfluorene data.

hydrocarbons, the modified MB is significantly better than the original method and is about as good as the modified Riedel, while the Lee-Kesler correlation is a close third.

EXPERIMENTAL DATA ON COAL LIQUID FRACTIONS

Vapor pressure data for six EDS fractions were presented by Wilson et al. (1981), who found, as in the case of the model compounds, that the modified Riedel equation was 50% better than the MB correlation in the superatmospheric region.

The analysis has been expanded here by including subatmospheric data, as well as data for SRC-II fractions (Gray, 1981; Gray and Holder, 1982; Gray et al., 1983). The boiling points, specific gravities, and Watson

Table 4.3   Characterization of Narrow Coal Liquid Fractions

Fraction		$t_b(°F)$	S	K_w	M
EDS[a]	S-1	406	0.915	10.42	134
EDS	S-2	549	1.005	9.98	179
EDS	U-1	473	0.962	10.16	158
EDS	U-2	655	1.059	9.79	225
EDS	N-1	458	0.954	10.19	164
EDS	N-2	655	1.036	10.01	205
SRC-II[b]	4HC-A	277	0.816	11.07	110[d]
SRC-II	6HC	382	0.951	9.93	127
SRC-II	7HC-B	427	0.967	9.93	140
SRC-II	10HC-B	570	1.002	10.08	188
SRC-II	15HC-B	678	1.083	9.64	218
SRC-II	18HC-B	875	1.166	9.44	285
SRC-II[c]	5-HC	320	0.8827	10.43	116
SRC-II	8-HC	476	0.9718	10.06	158
SRC-II	11-HC	643	1.0359	9.97	212
SRC-II	16-HC	726	1.0910	9.72	237
SRC-II	17-HC	787	1.1204	9.67	258
SRC-II	19-HC-A	938	1.1792	9.41	315

[a]More complete information on EDS fractions can be found in Appendix A.

[b]SRC-II data taken from Gray and Holder (1982).

[c]SRC-II data taken from Gray (1981).

[d]Calculated values reported by Gray and Holder (1982) and by Gray (1981).

characterization factors for the EDS and SRC-II fractions are listed in Table 4.3. More extensive inspection data for the EDS fractions are given in Appendix A.

Table 4.4 summarizes the results of the evaluation of the four vapor pressure methods against data for EDS and SRC-II fractions. Perhaps the most interesting result is that the modified MB correlation does better than the modified Riedel equation, most likely because the need to estimate the t_c and P_c of fractions degrades the quality of the predictions with the Riedel equation.

Table 4.4   Vapor Pressure Predictions for Coal Liquid Fractions[a]

Fraction		No. Points	Maxwell-Bonnell	Modified Maxwell-Bonnell	Lee-Kesler	Modified Riedel
			Percent Average Deviation			
EDS	S-1	(15)	5.79	3.09	1.40	1.17
EDS	S-2	(17)	9.50	5.11	5.02	4.53
EDS	U-1	(17)	4.22	2.56	4.42	3.95
EDS	U-2	(17)	12.80	7.87	8.77	7.50
EDS	N-1	(16)	8.93	5.26	4.90	4.62
EDS	N-2	(14)	12.37	5.22	5.76	3.73
SRC-II	4HC-A	(14)	3.24	3.09	2.02	4.21
SRC-II	6HC[b]	(15)	7.80	11.16	12.40	11.84
SRC-II	7HC-B[b]	(14)	4.53	8.22	9.84	9.02
SRC-II	10HC-B	(13)	6.76	2.32	3.97	4.72
SRC-II	15HC-B	(13)	8.99	5.57	5.18	5.51
SRC-II	18HC-B	(8)	18.85	4.89	8.18	5.03
SRC-II	5-HC[b]	(14)	3.44	4.01	4.20	5.03
SRC-II	8-HC[b]	(16)	3.00	7.12	8.30	8.34
SRC-II	11-HC[b]	(12)	7.63	3.18	5.26	4.80
SRC-II	16-HC[b]	(8)	5.94	2.00	2.76	3.48
SRC-II	17-HC[b]	(6)	5.85	1.82	2.43	2.70
SRC-II	19-HC-A[b]	(5)	18.59	10.85	9.99	12.11
All fractions		(234)	7.69	5.23	5.85	5.63
Excluding 6HC & 7HC-B		(205)	7.89	4.60	5.09	4.94

[a]Vapor pressure data were taken from Wilson et al. (1981), for the EDS fractions, and from Gray (1981), Gray et al. (1982), and Gray and Holder (1982), for the SRC-II fractions.

[b]Not included in data base used in the modification of the MB correlation.

Another interesting result of the comparison in Table 4.4 is that, except for the original MB correlation, the predictions improve when the two SRC-II fractions rich in phenols, 6HC and 7HC-B (Gray and Holder, 1982), are excluded. This supports the observation made earlier for m-cresol, 2,4-xylenol, and the other heteroatom-containing compounds: neither the (t_b, S) nor the (t_c, P_c, ω) framework should be expected to be

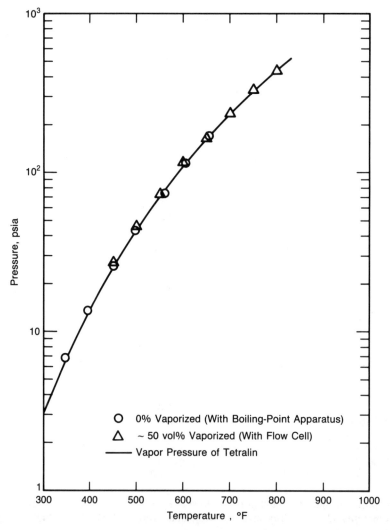

Figure 4.11 Total pressure measurements on EDS fraction S-1 versus vapor pressure of tetralin.

accurate for strongly polar compounds and fractions. This is further considered in Chapter 10.

The vapor pressure measurements on the fractions provide an excellent illustration of the similarity between narrow fractions and defined compounds. In Figure 4.11, the experimental measurements on the S-1 fraction (50 wt% distilled at 406°F) are superimposed on the vapor pressure of tetralin ($t_b = 405.7°F$). The good agreement strongly supports the assumption that a relatively narrow fraction behaves as a single component. This is also supported by the excellent agreement between the boiling point apparatus (at 0% vaporized) and flow cell measurements (at roughly 50 vol% vaporized); see Wilson et al. (1981).

The agreement shown in Figure 4.11 is corroborated in Figure 4.12, where the experimental measurements on the U-1 fraction (50 wt% distilled at 473°F) are superimposed on the vapor pressure of 1-methylnaphthalene ($t_b = 472.4°F$).

As far as vapor pressure measurement and prediction are concerned, it

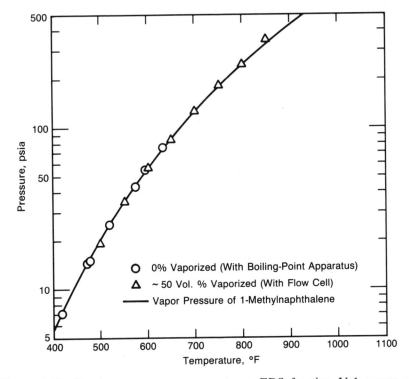

Figure 4.12 Total pressure measurements on EDS fraction U-1 versus vapor pressure of 1-methylnaphthalene.

appears that a fraction with $\Delta t_b \lesssim 100°F$ (between 1 and 99 wt% distilled) behaves much like a single defined compound. The equivalence is more satisfactory as the boiling point range decreases (it is better for S-1/tetralin than for U–1/1-methylnaphthalene), and clearly a fraction with $\Delta t_b \approx 50°F$ should be narrow enough to be treated as a single component.

CONCLUSIONS AND RECOMMENDATIONS

The Maxwell-Bonnell correlation for $K_w = 12.0$, Eqs. 4.1–4.2 or Figures 4.1–4.5, can continue to be the basis for the calculation of the vapor pressure of fractions. However, for coal-liquid fractions, that is, for $K_w <$ 12, the boiling point correction should be calculated with Eqs. 4.5–4.6 or Figure 4.9.

An alternative to MB for fractions is the Lee-Kesler (1980) correlation. Although this is better than the original MB, it is slightly inferior to the modified MB for aromatic compounds and fractions.

In the case of model compounds, the modified Riedel equation is clearly superior to the modified MB and Lee-Kesler correlations. However, the Riedel equation loses its advantage in the case of fractions, where the modified MB method is the best. This reversal is likely due to uncertainties in predicting t_c and P_c for the fractions. Thus if t_c (especially) and P_c are known experimentally or can be predicted reliably, the modified Riedel equation is the method of choice. Otherwise, the modified MB correlation should be preferred, especially since it does not require t_c, P_c, and ω.

One inherent weakness of the modified MB correlation, as well as of the Lee-Kesler correlation, is in handling heteroatom-containing compounds. That was more evident in the evaluation of the data for model compounds (Table 4.2), a consequence of their containing higher levels of heteroatoms than the two phenol-rich SRC-II fractions 6HC and 7HC-B (Table 4.4). Possible improvements in the vapor pressure correlations to make them applicable to heteroatom-containing compounds and fractions, especially those that are polar, are considered in Chapter 10.

In the next chapter, vapor pressure calculations are made with the modified MB correlation and the modified Riedel equation. When the modified MB correlation is used, as noted earlier, P_c does not have to be predicted with the methods described in Chapter 3. Instead, P_c is the vapor pressure calculated at t_c. Then this P_c and the vapor pressure at $T_r = 0.7$ are substituted into Eq. 4.13 to calculate ω, the acentric factor. On the other hand, when vapor pressure is calculated with the modified Riedel equation, P_c must be known, or predicted with Eq. 3.15, and ω must be calculated with Eq. 4.15.

REFERENCES

American Petroleum Institute, *Technical Data Book—Petroleum Refining*, 4th Ed., API, Washington, DC, 1983.

Andon, R. J. L., et al., Thermodynamic Properties of Organic Oxygen Compounds. Part 1. Preparation and Physical Properties of Pure Phenol, Cresols, and Xylenols, *J. Chem. Soc.*, 5246 (1960).

Boyd, R. H., S. N. Sanwal, S. Shary-Tehrany, and D. McNally, The Thermochemistry, Thermodynamic Functions, and Molecular Structures of Some Cyclic Hydrocarbons, *J. Phys. Chem.*, **75**, 1264 (1971).

Camin, D. L., and F. D. Rossini, Physical Properties of 14 API Research Hydrocarbons, C_9 to C_{15}, *J. Phys. Chem.*, **59**, 1173 (1955).

Chipman, J., and S. B. Peltier, Vapor Pressure and Heat of Vaporization of Diphenyl, *Ind. Eng. Chem.*, **21**, 1106 (1929).

Crafts, J. M. (1915); data reported by Timmermans (1950).

Garrick, F. J., The Vapour Pressures of Diphenyl and of Aniline, *Trans. Faraday Soc.*, **23**, 560 (1927).

Glaser, F., and H. Rüland, Investigations of the Vapor Pressures and Critical Data of Some Industrially Important Organic Substances, *Chem.-Ing.-Tech.*, **29**, 772 (1957); in German.

Gray, J. A., Selected Physical, Chemical, and Thermodynamic Properties of Narrow Boiling Range Coal Liquids from the SRC-II Process, Report No. DOE/ET/10104-7, April 1981.

Gray, J. A., C. J. Brady, J. R. Cunningham, J. R. Freeman, and G. M. Wilson, Thermophysical Properties of Coal Liquids. 1. Selected Physical, Chemical, and Thermodynamic Properties of Narrow Boiling Range Coal Liquids, *Ind. Eng. Chem. Process Des. Dev.*, **22**, 410 (1983).

Gray, J. A., and G. D. Holder, Selected Physical, Chemical, and Thermodynamic Properties of Narrow Boiling Range Coal Liquids from the SRC-II Process, Supplementary Property Data, Report No. DOE/ET/10104-44, April 1982.

Henson, B. J., A. R. Tarrer, C. W. Curtis, and J. A. Guin, Solubility of Carbon Dioxide and Methane in Coal-Derived Liquids at High Temperatures and Pressures, *Ind. Eng. Chem. Process Des. Dev.*, **21**, 575 (1982).

Herz, W., and P. Schuftan (1922); data reported by Timmermans (1950).

Jacquerod, A., and E. Wassmer (1904); data reported by AIChE DIPPR Project 801, first 193 compounds (1983).

Kara, M., S. Sung, G. E. Klinzing, and S. H. Chiang, Tetralin Vapour Pressure at Elevated Temperatures, *Fuel*, **60**, 633 (1981).

Kobayashi, R., Phase and Volumetric Equilibria in Coal Hydrogenation Systems, Quarterly Report to DOE, Sept. 1978.

Kudchadker, A. P., S. A. Kudchadker, and R. C. Wilhoit, *Anthracene and Phenanthrene*, API Publication 708, 1979.

Lee, B. I., and M. G. Kesler, Improve Vapor Pressure Prediction, *Hydrocarbon Process.*, **59**(8), 163 (1980).

Lenoir, J. M., K. E. Hayworth, and H. G. Hipkin, Enthalpies of Decalins and of *trans*-Decalin and *n*-Pentane Mixtures, *J. Chem. Eng. Data*, **16**, 129 (1971).

Malanowski, S., Vapour Pressures and Boiling Temperatures of Some Quinoline Bases, *Bull. Acad. Pol. Sci. Ser. Sci. Chim.*, **9**, 71 (1961).

Maxwell, J. B., and L. S. Bonnell, *Vapor Pressure Charts for Petroleum Hydrocarbons*, Exxon Research and Engineering Company, Florham Park, NJ, 1955; reprinted 1974.

——, Derivation and Precision of a New Vapor Pressure Correlation for Petroleum Hydrocarbons, *Ind. Eng. Chem.*, **49**, 1187 (1957).

Moor, G., and E. K. Kanep (1936); data reported by AIChE DIPPR Project 801, first 193 compounds (1983).

Mortimer, F. S., and R. V. Murphy, The Vapor Pressures of Some Substances Found in Coal Tar, *Ind. Eng. Chem.*, **15**, 1140 (1923).

Myers, H. S., and M. R. Fenske, Measurement and Correlation of Vapor Pressure Data for High Boiling Hydrocarbons, *Ind. Eng. Chem.*, **47**, 1652 (1955).

Nasir, P., S. C. Hwang, and R. Kobayashi, Development of an Apparatus to Measure Vapor Pressures at High Temperatures and Its Application to Three Higher-Boiling Compounds, *J. Chem. Eng. Data*, **25**, 298 (1980).

Nelson, O. A., and C. E. Senseman, Vapor Pressure Determinations on Naphthalene, Anthracene, Phenanthrene, and Anthraquinone between Their Melting and Boiling Points, *Ind. Eng. Chem.*, **14**, 58 (1922).

Riedel, L., A New Universal Vapor Pressure Formula, *Chem.-Ing.-Tech.*, **26**, 83 (1954); in German.

Sebastian, H. M., J. J. Simnick, H. M. Lin, and K. C. Chao, Gas-Liquid Equilibrium in Mixtures of Hydrogen and Quinoline, *J. Chem. Eng. Data*, **23**, 305 (1978a).

——, Gas-Liquid Equilibria in Mixtures of Hydrogen and Thianaphthene, *Can. J. Chem. Eng.*, **56**, 743 (1978b).

Simnick, J. J., C. C. Lawson, H. M. Lin, and K. C. Chao, Vapor-Liquid Equilibrium of Hydrogen/Tetralin System at Elevated Temperatures and Pressures, *AIChE J.* **23**, 469 (1977).

Simnick, J. J., K. D. Liu, H. M. Lin, and K. C. Chao, Gas-Liquid Equilibrium in Mixtures of Hydrogen and Diphenylmethane, *Ind. Eng. Chem. Process Des. Dev.*, **17**, 204 (1978).

Simnick, J. J., H. M. Sebastian, H. M. Lin, and K. C. Chao, Gas-Liquid Equilibrium in Mixtures of Hydrogen + *m*-Xylene and *m*-Cresol, *J. Chem. Thermodyn.*, **11**, 531 (1979).

Sivaraman, A., and R. Kobayashi, Investigation of Vapor Pressures and Heats of Vaporization of Condensed Aromatic Compounds at Elevated Temperatures, *J. Chem. Eng. Data*, **27**, 264 (1982).

Sivaraman, A., R. J. Martin, and R. Kobayashi, A Versatile Apparatus to Study the Vapor Pressure and Heats of Vaporization of Carbazole, 9-Fluorenone and 9-Hydroxyfluorene at Elevated Temperatures, *Fluid Phase Equilibria*, **12**, 175 (1983).

Stull, D. R., Vapor Pressure of Pure Substances: Organic Compounds, *Ind. Eng. Chem.*, **39**, 517 (1947).

TRCHP (Thermodynamic Research Center Hydrocarbon Project), Selected Values of Properties of Hydrocarbons and Related Compounds, Texas A & M U., College Station, sheets extant 1983.

Timmermans, J., *Physico-Chemical Constants of Pure Organic Compounds*, Elsevier, New York, 1950.

Ward, S. H., and M. Van Winkle, Vapor-Liquid Equilibria at 200 mm of Mercury, *Ind. Eng. Chem.*, **46**, 338 (1954).

Wieczorek, S. A., and R. Kobayashi, Vapor-Pressure Measurements of Diphenylmethane, Thianaphthene, and Bicyclohexyl at Elevated Temperatures, *J. Chem. Eng. Data*, **25**, 302 (1980).

————, Vapor-Pressure Measurements of 1-Methylnaphthalene, 2-Methylnaphthalene, and 9,10-Dihydrophenanthrene at Elevated Temperatures, *J. Chem. Eng. Data*, **26**, 8 (1981).

Wilson, G. M., R. H. Johnston, S. C. Hwang, and C. Tsonopoulos, Volatility of Coal Liquids at High Temperatures and Pressures, *Ind. Eng. Chem. Process Des. Dev.*, **20**, 94 (1981).

Yao, J., H. M. Sebastian, H. M. Lin, and K. C. Chao, Gas-Liquid Equilibria in Mixtures of Hydrogen and 1-Methylnaphthalene, *Fluid Phase Equilibria*, **1**, 293 (1977/78).

Zudkevitch, D., P. D. Krautheim, and D. Gaydos, Vapor Pressure of Coal-Liquid Fractions—Data and Correlation, *Fluid Phase Equilibria*, **14**, 117 (1983).

Chapter 5

VAPOR–LIQUID EQUILIBRIA

As noted in Chapter 2, uncertainties in vapor–liquid equilibria affect EDS and other direct coal liquefaction processes primarily in two areas: the liquefaction reactor (at ~850°F) and the vacuum pipestill (at ~700°F). At the low operating pressure of the vacuum pipestill, about 0.5 psia, no significant amounts of gases and light hydrocarbons are present, and therefore the volatility of coal liquids can be predicted with the vapor pressure methods discussed in Chapter 4.

At the high pressure of the liquefaction reactor, 2000–3000 psia, quantitative description of the VLE (vapor–liquid equilibrium) behavior requires methods that account for nonideal mixing in both phases and for the approach to the mixture critical point. Examples of such methods are the various cubic forms of the equation of state used in the industry for high-pressure VLE calculations. The method used in Chapter 5 is RKJZ, the Joffe-Zudkevitch modification of the Redlich-Kwong equation of state.

The high operating pressure of the reactor is primarily due to the partial pressure of H_2 (about 80%) and methane (about 10% of the total) Therefore, the volatility and solubility of these components in coal liquids are of special concern here. Indeed, most of the high-temperature VLE data examined in this chapter are on mixtures of model compounds or coal liquids with H_2 or methane. Measurements up to 880°F have been made possible by the use of flow cells (first by G. M. Wilson and then by K. C. Chao) that reduce the residence time to well under one minute, and thus minimize the effect of thermal decomposition.

After an introductory discussion on VLE calculations, the RKJZ method is described, and then this method is employed in the analysis of high-temperature VLE data on binaries of model compounds with H_2 or methane. Following the binary data, RKJZ is evaluated against data on ternary systems and, more extensively, on coal liquid mixtures. Although most of the available data are on EDS liquids, some information on SRC-II liquids is also included. Finally, there are concluding remarks and a recommendation for which of the two vapor pressure correlations selected in Chapter 4 should be used.

HIGH-PRESSURE VLE CALCULATIONS

High-pressure VLE calculations are usually carried out, at least in the industry, with various forms of the equation of state. One reason for the popularity of equations of state, in comparison to hybrid models such as the Chao-Seader correlation, is the great simplification they introduce in the calculations.

The starting point of all VLE calculations is the equilibrium condition:

$$f_i^V = f_i^L \tag{5.1}$$

That is, the fugacity of component i in the vapor should be equal to its fugacity in the liquid. If the fugacity is expressed in terms of the fugacity coefficient, Eq. 5.1 becomes

$$\phi_i^V y_i P = \phi_i^L x_i P \tag{5.2}$$

and the following expression for the K-value results:

$$K_i = \frac{y_i}{x_i} = \frac{\phi_i^L}{\phi_i^V} \tag{5.3}$$

Equation 5.3 provides the simplest and most direct method for VLE calculations—it requires no standard states or special procedures as the critical point is approached—provided the equation of state applies to both phases. Some equations do well—if we exclude strongly polar compounds and electrolytes. The industry today primarily uses cubic equations that are variations of the van der Waals equation of state.

Redlich and Kwong (1949) proposed a cubic equation that was significantly better than the van der Waals equation, but was still not satisfactory for VLE predictions. What was needed to improve VLE predictions was recommended by Wilson (1964), who suggested that the a parameter in the Redlich-Kwong equation (see Eq. 5.4) be made a function of temperature to improve the prediction of the pure-compound vapor pressure—and, hence, the prediction of multicomponent VLE as well.

Wilson's recommendation was largely ignored, even though he generalized the $a(T)$ function and applied his equation to many different systems, including those with H_2 (Wilson, 1966, 1969). It was not until Soave's (1972) modification that this approach gained popularity.

This chapter focuses on the Joffe-Zudkevitch modification of the Redlich-Kwong equation of state. A recent paper (Gray et al., 1983) examined RKJZ and compared it with Soave's modification of the Redlich-Kwong and the Peng-Robinson (1976) equation of state. The conclusion was that RKJZ was significantly better than the other two equations for H_2/aromatic compound mixtures.

A comparison of RKJZ with a modified Chao-Seader correlation was presented by Wilson et al. (1981). Again, RKJZ was significantly superior for binaries of aromatic compounds with either H_2 or methane, the major gas components in direct coal liquefaction.

Because of these earlier comparisons, only the RKJZ equation is examined in this chapter. The vapor pressure of coal-liquid fractions is calculated with two correlations: modified Maxwell-Bonnell or modified Riedel. On the other hand, the vapor pressure of model compounds has been largely based on the fit of experimental data with Eq. 4.14.

JOFFE-ZUDKEVITCH MODIFICATION OF THE REDLICH-KWONG EQUATION OF STATE

The Redlich-Kwong equation of state, in the form popularized by Soave (1972), is

$$P = \frac{RT}{v - b} - \frac{a(T)}{v(v + b)} \tag{5.4}$$

where

$$a = 0.42748 \cdots R^2 T_c^2 / P_c \cdot \alpha \tag{5.5}$$

$$b = 0.08664 \cdots RT_c / P_c \tag{5.6}$$

α introduces the temperature dependence in $a(T)$; in the original equation (Redlich and Kwong, 1949), it is given by

$$\alpha = 1/T_r^{0.5} \tag{5.7}$$

Wilson and Soave improved the vapor pressure predictions by making α a more complex function of T_r, as well as of the acentric factor.

In RKJZ (Zudkevitch and Joffe, 1970), both a and b in Eq. 5.4 are functions of temperature. In the form of this modification recommended by Joffe et al. (1970) and Gray (1977), the temperature dependence of a and b is determined by simultaneously matching liquid density and forcing the vapor and liquid fugacities to be equal at the pure component's vapor pressure. Above the critical temperature, a and b are set equal to their respective values at the critical temperature. (This makes the a and b *independent* of the critical constants for subcritical components, a definite advantage for high-boiling components.)

For H_2, which is above its critical temperature for all applications of interest, RKJZ uses Eq. 5.7 with the temperature-dependent criticals proposed by Chueh and Prausnitz (1967):

$$T_c = 78.5/(1 + 19.47/T) \tag{5.8}$$

$$P_c = 296.9/(1 + 39.47/T) \tag{5.9}$$

$$\omega = 0 \tag{5.10}$$

where T is in °R and P in psia.

Mixing Rules

The mixing rules in the cubic equations of state provide formulas for predicting the a and b of mixtures. The most common or "classical" mixing rules are the one-fluid van der Waals mixing rules:

$$a_m = \sum_i \sum_j z_i z_j a_{ij} \tag{5.11}$$

$$a_{ij} = (a_i a_j)^{0.5}(1 - C_{ij}) \tag{5.12}$$

$$b_m = \sum_i z_i b_i \tag{5.13}$$

$C_{ij} = 0$ is a reasonable assumption for most hydrocarbon/hydrocarbon binaries, especially at or above ambient temperature. However, nonzero C_{ij}'s are required when the components are very different in molecular size; for example, for H_2/hydrocarbon mixtures. Nonzero C_{ij}'s may also be required at very low temperatures, for calculations in the critical region, or for polar components; for example, water with hydrocarbons.

Although the use of only one binary constant has proven, in most cases, satisfactory, a second binary constant can be introduced by replacing Eq. 5.13 with

$$b_m = \sum_i \sum_j z_i z_j b_{ij} \tag{5.14}$$

$$b_{ij} = 0.5(b_i + b_j)(1 + D_{ij}) \tag{5.15}$$

Setting $D_{ij} = 0$ in Eq. 5.15 gives us the original mixing rule, Eq. 5.13. A nonzero D_{ij} has been found to help in fitting VLE data for CO_2 binaries (Turek et al., 1980) and H_2 binaries (Mathias and Stein, 1983).

The K-value of component i is given by the ratio of its fugacity coefficients in the liquid and vapor phases (Eq. 5.3). When the Redlich-Kwong equation is used, the fugacity coefficient of component i in the mixture is given by

$$\ln \phi_i = \ln \frac{v}{v - b} + \frac{2 \sum_j z_j b_{ij} - b}{v - b} - \frac{2 \sum_j z_j a_{ij}}{RT^{1.5} b} \ln \frac{v + b}{v}$$

$$+ \frac{a(2 \sum_j z_j b_{ij} - b)}{RT^{1.5} b^2} \left(\ln \frac{v + b}{v} - \frac{b}{v + b} \right) - \ln \frac{Pv}{RT} \tag{5.16}$$

where a and b are given by Eqs. 5.11 and 5.14, respectively, and v is the molar volume of the mixture in the appropriate phase.

The first step in the data analysis is to determine the optimum C_{ij}'s (and D_{ij}'s) by regressing VLE data for binaries of model compounds with H_2 or methane. These optimum C_{ij}'s are then used to predict the VLE of coal liquids.

ANALYSIS OF BINARY VLE DATA

The criterion for the determination of the binary interaction parameters, C_{ij} in Eq. 5.12 and D_{ij} in Eq. 5.15, is the minimization of the RMS (root-mean-square) of the sum of the percent deviations in the calculated K-values at fixed experimental conditions. This criterion was used by Gray (1977) and is exactly equivalent to the procedure proposed by Paunovic et al. (1981), who minimize the deviation between calculated component vapor and liquid fugacities for a given system in equilibrium.

This criterion is generally preferred to matching bubble-point pressure for typical high-pressure VLE calculations because it weighs equally the distribution of the light and heavy components in the binary mixture. In addition, the procedure of comparing K-values calculated at (fixed) experimental compositions produces smaller errors and is less complex than comparing K-values calculated from an equilibrium flash. However, both procedures yield nearly identical values of C_{ij}, except in the critical region, where the equilibrium condition becomes more sensitive to the respective phase compositions. The analysis generally excludes VLE data near the critical region.

In analyzing a particular binary system, certain individual component K-values were excluded from the regression to avoid biasing the correlations with inaccurate data. This often involved simply excluding points for which the component level in either phase dropped below some minimum value, typically 0.5 mole percent.

The H_2/tetralin binary has been picked to illustrate the fit of available VLE data with RKJZ. Figure 5.1 is a plot of the K-value of H_2 and Figure 5.2 is a similar plot for tetralin. Data have been plotted for four isotherms from three sources (Simnick et al., 1977; Nasir et al., 1980/81; Harrison, 1981). Generally, there is good agreement between the three sources, although Simnick's data tend to be on the high side. Furthermore, as also shown in the figures, RKJZ fits the data very satisfactorily with $C_{ij} = 0.24$, which was obtained by analyzing the data of Simnick, Nasir, and of Berryman et al. (1981); see Table 5.1. Additional data on H_2/tetralin at 658 and 732°F, also in good agreement with the earlier sources, have been reported by Kara et al. (1983).

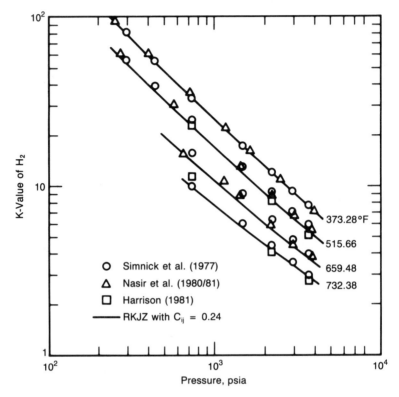

Figure 5.1 K-value of H_2 in the H_2/tetralin binary.

Table 5.1 summarizes the results of the data analyses for H_2 binaries, along with data sources, ranges of conditions, average errors, and optimum C_{ij}'s. Gray et al. (1983) evaluated the same data with the RKS (Redlich-Kwong-Soave) and PR (Peng-Robinson) equations of state, and RKJZ was found to have a significant advantage over the other methods.

Much of the advantage of the RKJZ method for binaries containing coal liquid model compounds lies in its previously noted independence from critical properties for subcritical components. This advantage is important because, for many of the model compounds, critical properties are not known accurately (see also Chapter 3), although vapor pressures are known in the temperature range of interest. Consequently, the RKS and PR methods will suffer in accuracy by comparison if the critical properties and acentric factor specified do not reproduce the vapor pressure in the region of interest. Gray et al. (1983) improved the fit of the RKS and PR methods by adjusting the critical properties to be more consistent with the vapor

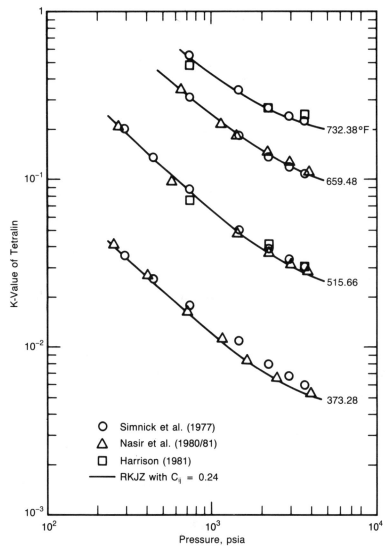

Figure 5.2 K-value of tetralin in the H_2/tetralin binary.

pressure data. However, some advantages still remained for the RKJZ method.

The fit with RKJZ of methane/model compound VLE data is summarized in Table 5.2. This table is taken from Wilson et al. (1981), who also used in their evaluation a modified Chao-Seader correlation. For both H_2 and methane binaries, RKJZ was significantly superior to the modified Chao-Seader correlation.

Table 5.1   Analysis of H_2/Model Compound VLE Data with RKJZ

Model Compound (No. Points)[a]	C_{ij} (Ave. Dev. in K-values)[b]	Data Sources
Tetralin (58/48)	0.240 (3.6/6.6)	Simnick et al. (1977); Nasir et al. (1980/81); Berryman et al. (1981)
1-Methylnaphthalene (36/36)	0.177 (3.4/4.9)	Yao et al. (1977/78); Lin et al. (1980a)
Cyclohexylcyclohexane (28/28)	0.410 (3.4/4.6)	Cukor and Prausnitz (1972); Sebastian et al. (1978c)
Diphenylmethane (27/27)	0.247 (1.9/4.7)	Simnick et al. (1978)
Phenanthrene (16/0)	0.224 (9.8/—)	DeVaney et al. (1978)
9,10-Dihydrophenanthrene (28/28)	0.719 (11.7/25.6)	Sebastian et al. (1979b)
m-Cresol (41/41)	0.302 (2.7/3.3)	Simnick et al. (1979a)
Quinoline (27/27)	0.265 (2.9/2.3)	Sebastian et al. (1978a)
Benzothiophene (27/27)	0.207 (3.4/2.3)	Sebastian et al. (1978b)
All compounds (288/262)	(4.3/6.6)	

[a] Number of points used in regression for H_2/model compound.
[b] Average absolute percent deviation in K-value for H_2/model compound.

Table 5.2   Analysis of Methane/Model Compound VLE Data with RKJZ

Model Compound (No. Points)[a]	C_{ij} (Ave. Dev. in K-values)[b]	Data Sources
Tetralin (24/24)	0.096 (3.5/4.2)	Sebastian et al. (1979a)
1-Methylnaphthalene (28/28)	0.09 (3.2/4.6)	Sebastian et al. (1979a)
Diphenylmethane (25/25)	0.08 (2.3/3.1)	Sebastian et al. (1979a)
m-Cresol (25/25)	0.12 (3.4/3.5)	Simnick et al. (1979b)
Quinoline (28/28)	0.10 (3.3/4.7)	Simnick et al. (1979c)
All compounds (130/130)	(3.1/4.0)	

[a] Number of points used in regression for methane/model compound.
[b] Average absolute percent deviation in K-value for methane/model compound.

Using Two Interaction Parameters (C_{ij} and D_{ij})

As mentioned earlier, the use of two interaction parameters has been investigated for improving VLE predictions, particularly for nonhydrocarbons in hydrocarbon systems [e.g., Turek et al. (1980) for CO_2/hydrocarbons]. Recently, Mathias and Stein (1983) applied PR to VLE data for H_2/coal liquid binaries by making use of two interaction parameters, with reasonably good results. However, C_{ij} was kept at a constant value of -1.0 for all binaries.

We have examined the use of two interaction parameters for improving VLE predictions over a wide temperature range, as an alternative to using temperature-dependent C_{ij}'s. We found that adding the second parameter does not appear to improve significantly the ability of cubic equations to represent the critical region, except in the case of hydrocarbon/water systems (Heidman et al., 1985).

However, as demonstrated in the following table, the use of two interaction parameters with RKJZ significantly improves its ability to predict both the low-temperature Henry's constant data of Cukor and Prausnitz (1972) and the high-temperature VLE data of Lin et al. (1980b) for H_2/hexadecane.

	Optimum Parameters		Average Absolute % Deviation in K-Value	
$T(K)$	C_{ij}	D_{ij}	H_2	Hexadecane
300–475	−0.33	0.0	1.0	—
461–664	0.122	0.0	8.0	5.3
300–664	−0.045	0.0	7.1	8.4
300–664	0.175	−0.025	1.8	2.6

Similar studies for a variety of other light gas binaries have shown that using two interaction parameters generally improves VLE predictions, especially for asymmetric mixtures of light gases with heavy paraffins. However, such an adjustment is often not necessary for mixtures of light gases with heavy aromatics. Moreover, adding a second constant may increase the difficulty of generalizing the interaction parameters for related systems.

Accordingly, to predict the VLE behavior of coal liquids, we have used only one binary constant, C_{ij}. For H_2/aromatic compound binaries, C_{ij} varied only between 0.18 and 0.30. This suggested that the value $C_{ij} = 0.24$ be used for all H_2/coal liquid fractions. (The very large value for H_2/9,10-dihydrophenanthrene, $C_{ij} = 0.719$, is suspect, especially since the optimum

C_{ij} value for the H_2 binary with cyclohexylcyclohexane, a naphthene, is 0.410.) It should also be noted that even a substantial change in C_{ij} has a minimal effect on the fit at the high temperatures of interest here. For instance, raising the RMS deviation for both K-values for H_2/tetralin by only 1% gave a band of C_{ij} values ±0.07 around the optimum ($C_{ij} = 0.24 \pm 0.07$).

Finally, the results on methane/aromatic binaries showed that, approximately, the value $C_{ij} = 0.10$ could be used for all methane/coal liquid fraction binaries.

Before we consider the VLE data on coal liquids, we examine briefly data on H_2-containing ternary systems.

HYDROGEN-CONTAINING TERNARY SYSTEMS

Generally, the mixing rules established for VLE calculations with equations of state neither require nor permit the use of ternary constants. Consequently, in this section we consider the ability of RKJZ to predict ternary VLE behavior for H_2-containing ternary systems, given good estimates of the binary interaction parameters.

Hydrogen/Methane/Tetralin Ternary

The H_2/methane/tetralin ternary system has been investigated by Simnick et al. (1980). RKJZ predicted the three K-values with an average absolute deviation of 4.7% (H_2), 3.2% (methane), and 6.6% (tetralin).

The experimental results shown in Figure 5.3, which also includes binary data (Simnick et al., 1977; Sebastian et al., 1979a; Nasir et al., 1980/81), indicate that, especially at high pressures, the tetralin K-values vary substantially with the vapor composition and reach a minimum at low methane concentrations. However, neither RKJZ nor the other cubic equations examined by Gray et al. (1983) predict such a minimum. It is likely that the tetralin K-values in the H_2/tetralin binary data reported by Simnick et al. (1977) are too high; see also Figure 5.2. Thus, better data may be needed to establish the composition dependence of the tetralin K-value.

Hydrogen/Tetralin/Diphenylmethane Ternary

The H_2/tetralin/diphenylmethane ternary provides a good test for the ability of RKJZ and other cubic equations of state to predict the volatility of H_2 in a heavy solvent mixture. Again, RKJZ predicted the data of Oliphant et al. (1979) very satisfactorily; the average absolute deviations in K-value were 5.5% (H_2), 4.4% (tetralin), and 4.2% (diphenylmethane).

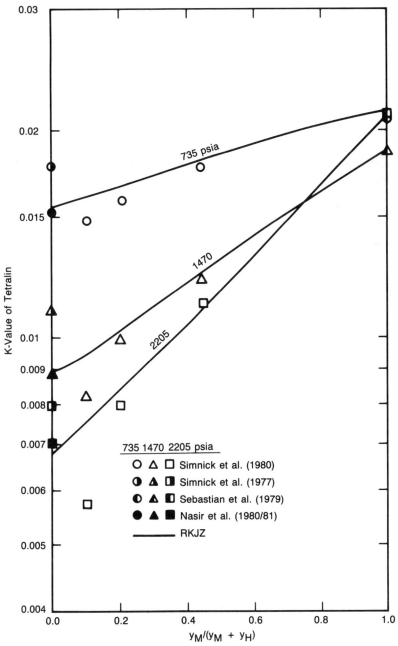

Figure 5.3 K-value of tetralin at 373°F in the ternary system H₂/methane/ tetralin (H: hydrogen; M: methane).

This limited comparison demonstrates that RKJZ can predict reliably the VLE of ternary systems with only one constant per binary, obtained from the fit of the binary data.

VLE DATA ON EDS COAL LIQUIDS

The three Illinois No. 6 coal liquids employed in our VLE work were identified in Figure 3.2. The boiling point versus weight percent distilled curves were determined by gas-chromatographic distillation, as illustrated in Figure 3.1. Liquids I and II are 400–750°F cuts, while Liquid III is a 400–1000°F cut. In addition, VLE measurements were made on two Wyoming Wyodak coal liquids, A-5 and A-6, that are similar in boiling range to Illinois Coal Liquids I and II. Table 5.3 lists the boiling-point distribution, specific gravity, and molecular weight for all five liquids. More extensive inspection data are given in Appendix A.

In the VLE calculations, the five coal liquids were represented by about 10 fractions. Boiling points and specific gravities for these fractions are given in Appendix B. With this information, the methods recommended in Chapters 3 and 4 can be used to calculate the molecular weight, t_c, P_c, ω, and the vapor pressure of each fraction.

Table 5.3   Boiling-Point Distribution, Specific Gravity, and Molecular Weight of Coal Liquids

wt% Distilled	Illinois Coal Liquids[a]			Wyoming Coal Liquids	
	I	II	III	WA-5	WA-6
	Boiling-Point Distribution (°F)				
1	387	356	367	350	350
10	407	389	404	390	390
30	450	432	477	430	432
50	504	483	580	478	488
70	570	552	738	545	563
90	675	668	912	695	658
99	831	800	1045	805	735
Specific gravity (60/60°F)	0.972	0.965	1.055	0.970	0.987
Molecular weight	168	169	193	167	169

[a]The values reported here supersede those reported by Wilson et al. (1981) and Hwang et al. (1983).

Table 5.4   RKJZ Binary Interaction
Constants C_{ij}

i	j	C_{ij}
H_2	CH_4	−0.039
H_2	H_2S	−0.04
H_2	Fraction	0.24
CH_4	H_2S	0.08
CH_4	Fraction	0.10
H_2S	Fraction	0.04
Fraction	Fraction	0.0

In addition to vapor pressure, RKJZ requires liquid density to establish the temperature dependence of its parameters a and b. Liquid density is discussed in Chapter 7.

The VLE data on mixtures of the five EDS coal liquids with H_2 and methane are given by Wilson et al. (1981) and Hwang et al. (1983). H_2S was also present for two runs (700°F and 1500 psia; 850°F and 1950 psia) with Illinois Coal Liquid I. With the three Illinois coal liquids, the H_2 level in the charge varied between 0.8 and 1.0 wt%, while the methane/H_2 weight ratio was close to 1.4, except for the two runs with H_2S, where the ratio was approximately 4.0. In the case of the Wyoming coal liquids, the H_2 level in the charge varied between 0.5 and 1.2 wt%, while the methane/H_2 weight ratio was 1.23.

The C_{ij}'s used in the calculations with RKJZ are listed in Table 5.4. Those for the key H_2/fraction and methane/fraction binaries were based on the analysis of data on model compound binaries (see Tables 5.1 and 5.2). The other C_{ij}'s were taken from previous work (Gray et al., 1983) or were determined by fitting literature VLE data.

Volatility of Coal Liquids

The volatility of coal liquids has been expressed in terms of the overall weight fraction vaporized. Since H_2 and methane have much lower molecular weights than the coal liquids, the overall weight fraction vaporized essentially represents only the vaporization of the coal liquids.

Figure 5.4 presents the measurements on Illinois Coal Liquid I (in the absence of H_2S). The corner in the 850°F isotherm is due to the variable H_2 level in the feed: 0.94 (1500 psia), 1.13 (1950 psia), and 0.78 wt% (2407 psia).

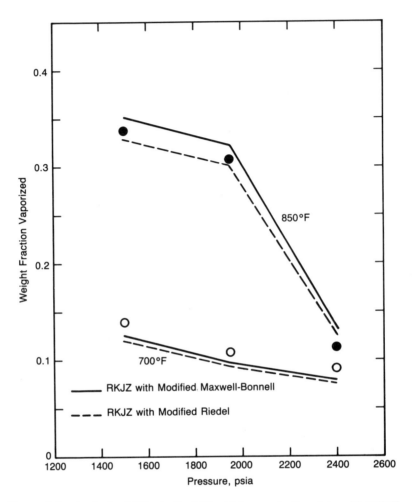

Figure 5.4   Analysis of Illinois Coal Liquid I vaporization data with RKJZ.

As shown in Figure 5.4, RKJZ underpredicts the data at 700°F by about 10%. At 850°F, RKJZ with the modified MB correlation overpredicts the data by about 10%, while with the modified Riedel equation the average deviation is less than 5%.

Figure 5.5, with the data for the heavier Coal Liquid III, shows the same trends as Figure 5.4: the modified Riedel equation gives slightly better results (average deviation of 3.4% versus 4.6% with the modified MB correlation). Even at 880°F, the highest temperature at which VLE measurements have been made, predictions are within 2% of the experimental data.

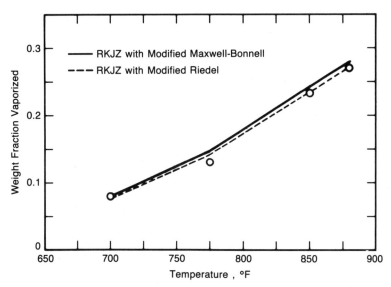

Figure 5.5 Analysis of Illinois Coal Liquid III vaporization data ($P = 1950$ psia) with RKJZ.

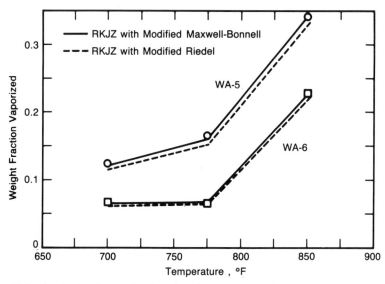

Figure 5.6 Analysis of vaporization data ($P = 1950$ psia) for Wyoming coal liquids with RKJZ.

Figure 5.6, with the vaporization data for the two Wyoming coal liquids, shows that RKJZ underpredicts the vaporization, and thus the modified MB correlation gives a lower average deviation than the modified Riedel: 2.4 vs. 5.0%. Since some of the vaporization data for Illinois coal liquids were also underpredicted, no clear distinction can be drawn between Illinois and Wyoming coal liquids.

A summary of the predictions of the weight fraction vaporized or the volatility of the coal liquids is given in Table 5.5. For all 24 points on the five liquids, the RKJZ/modified MB combination is slightly better than RKJZ/Riedel: 8.91 versus 9.43% average deviation in the predicted weight fraction vaporized. Hwang et al. (1983) have shown that RKJZ with the original MB gives an average deviation of 11.9%, while the MB/Riedel combination gives 11.5%, which also is the result with the modified Chao-Seader method when the critical constants are calculated with the methods recommended in Chapter 3. Thus 9% average deviation appears to be the best that can be done with the VLE data on the EDS coal liquids. More data on well-characterized coal liquids are needed to establish which of the vapor pressure methods is better or whether the source of coal makes a difference.

Table 5.5   Prediction of Volatility of EDS Coal Liquids with RKJZ[a]

	Vapor Pressure with			
	Modified Maxwell-Bonnell		Modified Riedel	
	% AAD	% Bias	% AAD	% Bias
Illinois Coal Liquids				
18 points	11.08	−0.51	10.90	−4.99
17 points[b]	9.31	+2.39	8.88	−2.62
Wyoming Coal Liquids				
6 points	2.39	−1.55	5.03	−5.03
Total				
24 points	8.91	−0.77	9.43	−5.00
23 points[b]	7.50	+1.36	7.88	−3.25

[a] % dev = $100 \times (\text{calc} - \text{exp})/\text{exp}$; % AAD = $(1/N) \sum |(\% \text{ dev})_i|$;
% bias = $(1/N) \sum (\% \text{ dev})_i$; N = number of data points.

[b] Removed one point for which deviations are 41–45%.

K-Values of Light Components in Coal Liquids

Light components such as H_2 are much more volatile than the coal-liquid components. A sample plot of all the K-values, at 850°F and 1950 psia, is given in Figure 5.7. The K-values of H_2 and methane are of special interest here because they provide a link with the measurements on model compounds.

An illustration of the K-value data predictions for H_2 is given in Figure 5.8. The agreement of RKJZ/Riedel (which is somewhat better than RKJZ/modified MB) with the data deteriorates with increasing temperature. A very similar behavior is shown in Figure 5.9 for methane. The experimental K-values of H_2 and methane in WA-6 at 775°F are probably in error since they are higher than those in WA-5 by a factor of 1.6.

A summary of the predictions of the K-values of H_2 and methane in EDS coal liquids is given in Tables 5.6 and 5.7, respectively. Results with RKJZ/Riedel are slightly better than those with RKJZ/modified MB, but both procedures substantially underpredict the K-values.

The two measurements involving H_2S are insufficient to test the C_{ij} values used for H_2S binaries in the RKJZ calculations. However, as in the

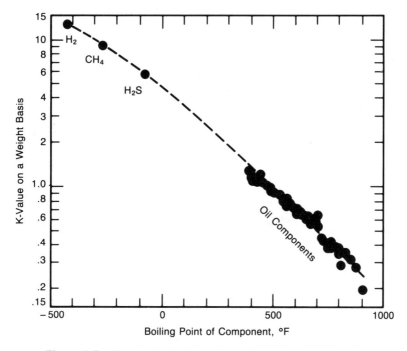

Figure 5.7 Sample plot of K-values at 850°F and 1950 psia.

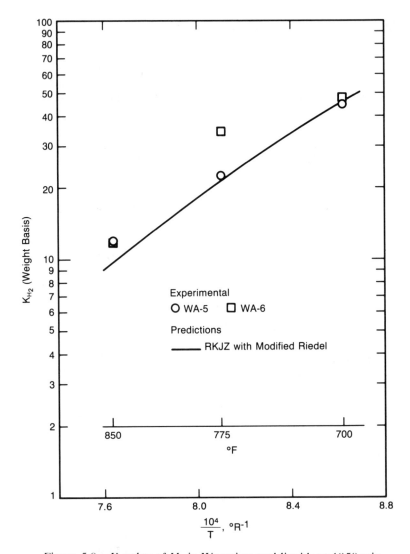

Figure 5.8 K-value of H_2 in Wyoming coal liquids at 1950 psia.

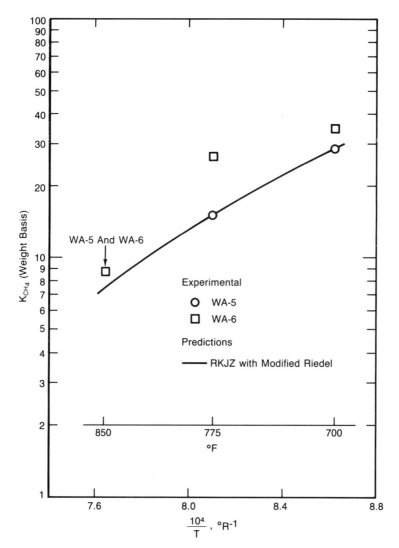

Figure 5.9 *K*-value of methane in Wyoming coal liquids at 1950 psia.

Table 5.6   Prediction of H_2 K-Value with RKJZ[a]

| | Vapor Pressure with | | | |
| | Modified Maxwell-Bonnell | | Modified Riedel | |
	% AAD	% Bias	% AAD	% Bias
Illinois Coal Liquids				
18 points	23.19	−21.97	22.17	−19.18
Wyoming Coal Liquids				
6 points	20.32	−20.32	17.14	−17.14
Total				
24 points	22.47	−21.56	20.91	−18.67

[a] See Table 5.5.

Table 5.7   Prediction of Methane K-Value with RKJZ[a]

| | Vapor Pressure with | | | |
| | Modified Maxwell-Bonnell | | Modified Riedel | |
	% AAD	% Bias	% AAD	% Bias
Illinois Coal Liquids				
18 points	19.70	−19.31	18.45	−16.53
Wyoming Coal Liquids				
6 points	20.62	−20.62	17.67	−17.67
Total				
24 points	19.93	−19.64	18.26	−16.82

[a] See Table 5.5.

case of H_2 and methane, RKJZ underpredicts the K-value (on a weight basis) of H_2S by 2% at 700°F and 19% at 850°F.

Figure 5.10 demonstrates the connection between VLE measurements on tetralin and 1-methylnaphthalene, on one hand, and those on a hydrogenated coal liquid such as I (modeled by tetralin) and an unhydrogenated cut such as III (modeled by 1-methylnaphthalene). The volatility of H_2 in

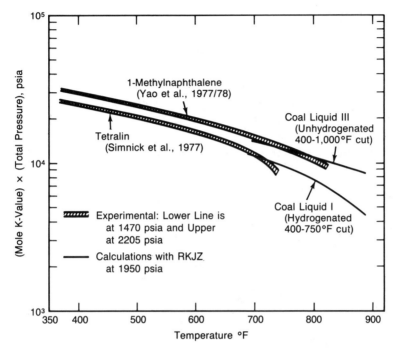

Figure 5.10   Volatility of H_2 in hydrocarbons and coal liquids.

coal liquids and the effect of hydrogenation on the volatility are very similar to those in the model compounds. Thus Figure 5.10 underlies the usefulness of model compounds and measurements made on them.

Effect of H_2 on Volatility of Coal Liquids

The effect of H_2 on the volatility of coal liquids is shown in Figure 5.11. Experimental data on weight fraction vaporized for Illinois Coal Liquids I and II, and for Wyoming Coal Liquids A-5 and A-6, are plotted as a function of weight percent H_2 charged. As mentioned earlier, the boiling-point ranges for these four coal liquids are very similar; roughly, they are 400–750°F cuts. The curves are the calculated volatilities of Illinois Coal Liquid II with RKJZ and the modified Riedel equation. (This comparison suggests that the experimental result on Illinois Coal Liquid II at 850°F with 0.855 wt% H_2 charged is most likely in error; this is the point excluded in Table 5.5.) Figure 5.11 also indicates that the coal source apparently has little effect on the volatility or VLE behavior of coal liquids, provided the

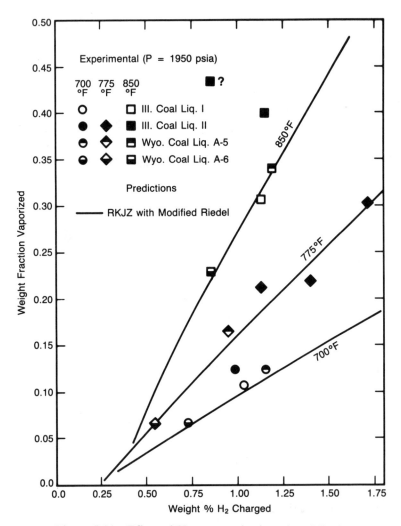

Figure 5.11   Effect of H₂ on vaporization of coal liquids.

liquids have a similar boiling-point distribution. Similarly, no differences are suggested between WA-5, which is unhydrogenated, and WA-6, which is hydrogenated.

LITERATURE VLE DATA ON COAL LIQUIDS

Recent investigations by Lin et al. (1981) and Kara et al. (1983) provide most of the available literature data on VLE and solubility of H_2 and methane in well-characterized EDS and SRC-II coal liquids. Other investigations include those by Prather et al. (1977) on the solubility of H_2 in creosote oil and in SRC recycle solvent, and Henson et al. (1982) on the solubility of CO_2 and methane, also in a creosote oil and an SRC recycle solvent. These latter investigators provide only partial analyses of the coal liquids and, therefore, only qualitative comparisons between predictions and data can be made. Finally, Harrison et al. (1981) reported multicomponent VLE data that included liquids from the H-Coal process. Unfortunately, these liquids are not adequately characterized to facilitate a detailed comparison with predictive calculations.

Table 5.8 summarizes comparisons of the solubility data reported by Lin et al. (1981) for H_2 and methane in EDS and SRC-II liquids with predictions obtained from the RKJZ equation (with the modified Riedel equation). Although the data reported by Lin et al. are at temperatures lower than those investigated by Wilson et al. (1981) and Hwang et al. (1983) for wide-boiling EDS liquids, Table 5.8 shows that RKJZ accurately predicts the solubility of H_2 and methane in the relatively narrow-boiling EDS and SRC-II liquids. Overall, average errors in the predicted H_2 solubilities are

Table 5.8   Summary of Deviations Between RKJZ Predictions and Data of Lin et al. (1981) for Solubility of H_2 and Methane in Coal Liquids[a]

Coal Liquid	No. of Points, H_2/Methane	% AAD and Bias in Predicted Solubilities[b]	
		H_2	Methane
EDS CLPP-A5 (400–500°F)[c]	10/10	4.66 (4.02)	3.13 (−1.31)
EDS CLPP-A5 (500–600°F)[c]	8/9	5.08 (−4.51)	7.78 (−7.78)
SRC-II No. 5	8/4	3.59 (−2.63)	7.21 (−5.85)
SRC-II No. 9	8/4	8.29 (−8.29)	6.61 (−6.61)
SRC-II No. 12	5/—	3.04 (−3.04)	—
Overall	39/27	5.06 (−2.52)	5.80 (−4.92)
EDS	18/19	4.85 (0.23)	5.33 (−4.37)
SRC-II	21/8	5.25 (−4.88)	6.91 (−6.23)

[a]Vapor pressures calculated with the modified Riedel equation.
[b]See Table 5.5.
[c]Boiling point range.

about 5% for both liquids, while errors in the predicted methane solubilities are about 5% for EDS liquids and 7% for the SRC-II liquids.

In addition, these comparisons show that RKJZ generally underpredicts the gas solubility and, therefore, overpredicts the volatility, that is, K-value, of the light components in the coal liquids. This is contrary to the results obtained from comparisons with the VLE data reported by Hwang et al. (1983). For those higher temperature data, Tables 5.6 and 5.7 show that RKJZ underpredicts the K-values of H_2 and methane in EDS liquids by about 20%.

Figure 5.12 illustrates the temperature and pressure dependence of the solubility of H_2 in the SRC-II No. 9 liquid, as reported by Lin et al. (1981). Also shown are solubility data for H_2 in both an SRC recycle solvent, as reported by Prather et al. (1977), and in an SRC-II hydrotreated solvent, as reported by Kara et al. (1983). The comparison in Figure 5.12 is consistent with the characterization data provided by Lin et al. and Kara et al., which suggest that the solubilities in the SRC-II No. 9 liquid and the hydrotreated solvent should be similar at comparable conditions.

Although no characterization data were reported by Prather et al. for their SRC recycle solvent, one might expect their H_2 solubilities to be similar to those reported by Lin et al. and Kara et al. However, Figure 5.12

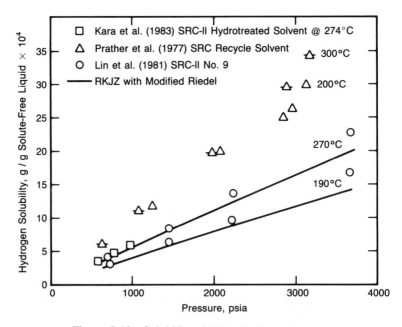

Figure 5.12   Solubility of H_2 in SRC coal liquids.

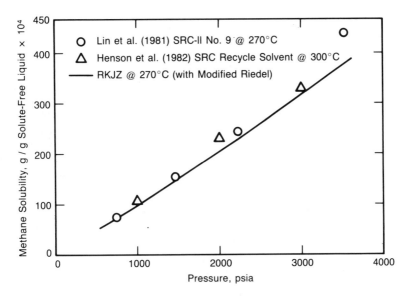

Figure 5.13   Solubility of methane in SRC coal liquids.

shows that the H_2 solubilities reported by Prather et al. are higher by a factor of 2 at comparable conditions. This suggests that either their SRC solvent has a substantially different characterization or their data are in error.

Figure 5.13 illustrates a similar comparison between data reported by Lin et al. and Henson et al. (1982) for the solubility of methane in SRC-II No. 9 and SRC recycle solvent, respectively. Henson et al. do not provide the characterization data needed for direct comparison with predictions by RKJZ, yet Figure 5.13 shows that their solubility data are in good agreement with those of Lin et al., and that both sets of data are reasonably well represented by RKJZ.

CONCLUSIONS AND RECOMMENDATIONS

One significant result of our VLE work (Wilson et al., 1981; Hwang et al., 1983) is that the volatility of coal liquids is now known experimentally at liquefaction reactor conditions, which are the extreme conditions in the process (about 850°F and 2000–2500 psia in the case of the EDS process). The experimental data on both Illinois and Wyoming coal liquids indicate that the coal source apparently has little effect on the volatility or VLE behavior of coal liquids, provided the coal liquids have a similar boiling-point distribution. There also seems to be little dependence of the volatility on the extent of hydrotreating. Furthermore, this volatility (expressed as the weight fraction vaporized) can be predicted with an average deviation of about 9%. Such an uncertainty should cause no insurmountable problems in the design of the EDS or other direct coal liquefaction processes.

The two vapor pressure correlations used in the VLE calculations produced similar results. The modified Maxwell-Bonnell gave slightly better results for the fraction vaporized. In the case of the K-values for H_2 and methane, the predictions were poor, but the Riedel equation was better in every case.

The analysis of literature data on the solubility of H_2 or methane in EDS and SRC-II coal liquids demonstrated that the RKJZ/Riedel predictions were very satisfactory for both liquids, and thus the process makes little difference. However, most of these data were at lower temperatures and on narrower cuts than in the work on EDS coal liquids (Wilson et al., 1981; Hwang et al., 1983). Although the high-temperature K-values for H_2 and methane in wide-boiling EDS liquids may be in error, at present we can only conclude that our predictions for these liquids at around 850°F are unsatisfactory, as they are as much as 50% low.

REFERENCES

Berryman, J. M., et al., Vapor-Liquid Equilibrium for H_2/Hydrocarbon/Nonhydrocarbon Systems, Final Report to API, 1981.

Chueh, P. L., and J. M. Prausnitz, Vapor-Liquid Equilibria at High Pressures. Vapor-Phase Fugacity Coefficients in Nonpolar and Quantum-Gas Mixtures, *Ind. Eng. Chem. Fundam.*, **6**, 492 (1967).

Cukor, P. M., and J. M. Prausnitz, Solubilities of Gases in Liquids at Elevated Temperatures. Henry's Constants for Hydrogen, Methane, and Ethane in Hexadecane, Bicyclohexyl, and Diphenylmethane, *J. Phys. Chem.*, **76**, 598 (1972).

DeVaney, W. E., J. M. Berryman, P. L. Kao, and B. E. Eakin, *High Temperature VLE Measurements for Substitute Gas Components*. GPA Research Report 30, February 1978.

Gray, R. D., Jr., Correlation of H_2/Hydrocarbon VLE Using Redlich-Kwong Variants, presented at the 1977 Annual AIChE Meeting, New York, November 13–17, 1977.

Gray, R. D., Jr., J. L. Heidman, S. C. Hwang, and C. Tsonopoulos, Industrial Applications of Cubic Equations of State for VLE Calculations, with Emphasis on H_2 Systems, *Fluid Phase Equilibria*, **13**, 59 (1983).

Harrison, R. H. (NIPER), private communication (1981).

Heidman, J. L., C. Tsonopoulos, C. J. Brady, and G. M. Wilson, High-Temperature Mutual Solubilities of Hydrocarbons and Water. II. Ethylbenzene, Ethylcyclohexane, and *n*-Octane, *AIChE J.*, **31**, 376 (1985).

Henson, B. J., A. R. Tarrer, C. W. Curtis, and J. A. Guin, Solubility of Carbon Dioxide and Methane in Coal Derived Liquids at High Temperatures and Pressures, *Ind. Eng. Chem. Process Des. Dev.*, **21**, 575 (1982).

Hwang, S. C., G. M. Wilson, and C. Tsonopoulos, Volatility of Wyoming Coal Liquids at High Temperatures and Pressures, *Ind. Eng. Chem. Process Des. Dev.*, **22**, 636 (1983).

Joffe, J., G. M. Schroeder, and D. Zudkevitch, Vapor-Liquid Equilibria with the Redlich-Kwong Equation of State, *AIChE J.*, **16**, 496 (1970).

Kara, M., et al., Vapor-Liquid Equilibrium in Hydrogen/Tetralin and Hydrogen/Coal-Derived Liquids, in S. A. Newman (ed.), *Chemical Engineering Thermodynamics*, Ann Arbor Science, Ann Arbor, MI, 1983, pp. 131–140.

Lin, H. M., H. M. Sebastian, and K. C. Chao, Gas-Liquid Equilibria of Hydrogen + 1-Methylnaphthalene at 457°C, *Fluid Phase Equilibria*, **4**, 321 (1980a).

————, Gas-Liquid Equilibrium in Hydrogen + n-Hexadecane and Methane + n-Hexadecane at Elevated Temperatures and Pressures, *J. Chem. Eng. Data*, **25**, 252 (1980b).

Lin, H. M., H. M. Sebastian, J. J. Simnick, and K. C. Chao, Solubilities of Hydrogen and Methane in Coal Liquids, *Ind. Eng. Chem. Process Des. Dev.*, **20**, 253 (1981).

Mathias, P. M., and F. P. Stein, Phase Equilibrium Studies, Report No. DOE/OR/03054/67, September 1983.

Nasir, P., R. J. Martin, and R. Kobayashi, A Novel Apparatus for the Measurement of the Phase and Volumetric Behavior at High Temperatures and Pressures and Its Application to Study VLE in the Hydrogen-Tetralin System, *Fluid Phase Equilibria*, **5**, 279 (1980/81).

Oliphant, J. L., H. M. Lin, and K. C. Chao, Gas-Liquid Equilibrium in Hydrogen + Tetralin + Diphenylmethane and Hydrogen + Tetralin + m-Xylene, *Fluid Phase Equilibria*, **3**, 35 (1979).

Paunovic, R., S. Jovanovic, and A. Mihajlov, Rapid Computation of Binary Interaction Coefficients of an Equation of State for Vapor-Liquid Equilibrium Calculations. Application to the Redlich-Kwong-Soave Equation of State, *Fluid Phase Equilibria*, **6**, 141 (1981).

Peng, D.-Y., and D. B. Robinson, A New Two-Constant Equation of State, *Ind. Eng. Chem. Fundam.*, **15**, 59 (1976).

Prather, J. W., et al., Solubility of Hydrogen in Creosote Oil at High Temperatures and Pressures, *Ind. Eng. Chem. Process Des. Dev.*, **16**, 267 (1977).

Redlich, O., and J. N. S. Kwong, On the Thermodynamics of Solutions. V. An Equation of State: Fugacities of Gaseous Solutions, *Chem. Rev.*, **44**, 233 (1949).

Sebastian, H. M., J. J. Simnick, H. M. Lin, and K. C. Chao, Gas-Liquid Equilibrium in Mixtures of Hydrogen and Quinoline, *J. Chem. Eng. Data*, **23**, 305 (1978a).

————, Gas-Liquid Equilibria in Mixtures of Hydrogen and Thianaphthene, *Can. J. Chem. Eng.*, **56**, 743 (1978b).

————, Gas-Liquid Equilibrium in Binary Mixtures of Methane with Tetralin, Diphenylmethane, and 1-Methylnaphthalene, *J. Chem. Eng. Data*, **24**, 149 (1979a).

————, Vapor-Liquid Equilibria in Hydrogen + 9,10-Dihydrophenanthrene Mixtures, *J. Chem. Eng. Data*, **24**, 343 (1979b).

Sebastian, H. M., J. Yao, H. M. Lin, and K. C. Chao, Gas-Liquid Equilibrium of the Hydrogen/Bicyclohexyl System at Elevated Temperatures and Pressures, *J. Chem. Eng. Data*, **23**, 167 (1978c).

Simnick, J. J., C. C. Lawson, H. M. Lin, and K. C. Chao, Vapor-Liquid Equilibrium of Hydrogen/Tetralin System at Elevated Temperatures and Pressures, *AIChE J.*, **23**, 469 (1977).

Simnick, J. J., K. D. Liu, H. M. Lin, and K. C. Chao, Gas-Liquid Equilibrium in Mixtures of Hydrogen and Diphenylmethane, *Ind. Eng. Chem. Process Des. Dev.*, **17**, 204 (1978).

Simnick, J. J., H. M. Sebastian, H. M. Lin, and K. C. Chao, Gas-Liquid Equilibrium in Mixtures of Hydrogen + *m*-Xylene and *m*-Cresol, *J. Chem. Thermodyn.*, **11**, 531 (1979a).

———, Gas-Liquid Equilibrium in Mixtures of Methane + *m*-Xylene, and Methane + *m*-Cresol, *Fluid Phase Equilibria*, **3**, 145 (1979b).

———, Vapor-Liquid Equilibrium in Methane + Quinoline Mixtures at Elevated Temperatures and Pressures, *J. Chem. Eng. Data*, **24**, 239 (1979c).

———, Vapor-Liquid Phase Equilibria in the Ternary System Hydrogen + Methane + Tetralin, *J. Chem. Eng. Data*, **25**, 147 (1980).

Soave, G., Equilibrium Constants from a Modified Redlich-Kwong Equation of State, *Chem. Eng. Sci.*, **27**, 1197 (1972).

Turek, E. A., R. S. Metcalfe, L. Yarborough, and R. L. Robinson, Jr., Phase Equilibria in Carbon Dioxide-Multicomponent Hydrocarbon Systems: Experimental Data and an Improved Prediction Technique, Soc. Petrol. Eng. AIME Paper 9231, 1980.

Wilson, G. M., Vapor-Liquid Equilibria Correlation by Means of a Modified Redlich-Kwong Equation of State, *Adv. Cryog. Eng.*, **9**, 168 (1964).

———, Calculation of Enthalpy Data from a Modified Redlich-Kwong Equation of State, *Adv. Cryog. Eng.*, **11**, 392 (1966).

———, A Modified Redlich-Kwong Equation of State; Application to General Physical Data Calculations, presented at the 1969 National AIChE Meeting, Cleveland, May 4–7, 1969.

Wilson, G. M., R. H. Johnston, S. C. Hwang, and C. Tsonopoulos, Volatility of Coal Liquids at High Temperatures and Pressures, *Ind. Eng. Chem. Process Des. Dev.*, **20**, 94 (1981).

Yao, J., H. M. Sebastian, H. M. Lin, and K. C. Chao, Gas-Liquid Equilibria in Mixtures of Hydrogen and 1-Methylnaphthalene, *Fluid Phase Equilibria*, **1**, 293 (1977/78).

Zudkevitch, D., and J. Joffe, Correlation and Prediction of Vapor-Liquid Equilibria with the Redlich-Kwong Equation of State, *AIChE J.*, **16**, 112 (1970).

Chapter 6

THERMAL PROPERTIES

This chapter examines the liquid heat capacity and, to a lesser extent, the heat of vaporization and heat of combustion of coal liquids. The vapor heat capacity, although very important in process calculations, is not considered extensively for two reasons. The first reason, as noted in the Preface and Chapter 1, is that the level of polynuclear aromatics in the vapor phase is generally very low. Therefore, the vapor enthalpy should be reasonably well predicted by correlations developed for petroleum liquids. The second reason is that the available vapor-phase enthalpy correlations appear to do well even for phenols, the most polar of the major constituents of coal liquids.

The basis for calculating thermal properties of coal liquids is the method developed by Kesler and Lee (1976) for predicting vapor and liquid enthalpies of petroleum-derived fluids. Their work improved upon the Johnson and Grayson (1961) correlation at high pressures and near the critical region. Kesler and Lee also developed the following correlation for the ideal-gas heat capacity (from 0 to 1200°F):

$$C_p^* = a + bT + cT^2 \tag{6.1a}$$

$$a = -0.32646 + 0.02678 K_w \\ - CF(0.084773 - 0.080809 S) \tag{6.1b}$$

$$b = -[1.3892 - 1.2122 K_w + 0.0383 K_w^2 \\ - CF(2.1773 - 2.0826 S)] \times 10^{-4} \tag{6.1c}$$

$$c = -[1.5393 + CF(0.78649 - 0.70423 S)] \times 10^{-7} \tag{6.1d}$$

$$CF = [(12.8/K_w - 1)(10.0/K_w - 1) \times 100]^2, \quad 10 \le K_w \le 12.8 \\ = 0, \qquad\qquad\qquad\qquad\qquad\quad \text{otherwise} \tag{6.1e}$$

[Although not stated explicitly by Kesler and Lee (1976), the correction term, CF, should be used only for K_w in the range from 10 to 12.8; otherwise, CF should equal zero.]

Equation 6.1 was compared with data for the model compounds 1-methylnaphthalene ($K_w = 9.536$) and *m*-cresol ($K_w = 9.14$) in order to

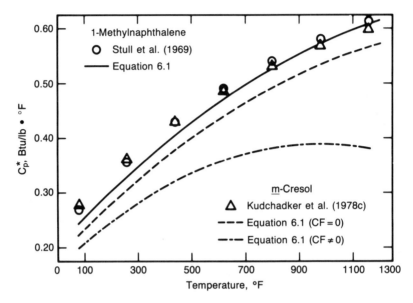

Figure 6.1   Ideal-gas heat capacity of coal liquid model compounds.

demonstrate its applicability to coal liquids. As shown in Figure 6.1, the predicted ideal-gas heat capacities agree remarkably well with the experimental data. Furthermore, Figure 6.1 shows that predictions would have been very poor (and would have the wrong temperature dependence) if, incorrectly, nonzero values of CF had been used. Thus we recommend that Eq. 6.1 be used for the ideal-gas heat capacity of coal liquids.

In the remainder of the chapter, we focus our attention on predicting the liquid heat capacity, heat of vaporization, and heat of combustion. Two widely used correlations for predicting liquid heat capacities of petroleum-derived liquids are those of Watson and Nelson (1933) and of the API Technical Data Book (1983). However, these correlations must be slightly modified to make them more suitable for coal liquids. We also examine the heat of vaporization of coal liquids and discuss the applicability of the generalized correlation (Lee and Kesler, 1975) included in the enthalpy-prediction method proposed by Kesler and Lee (1976). Finally, we examine several available correlations, as well as present a new one, for the heat of combustion of coal liquids. These correlations are typically used for predicting the heating values of coal-derived liquid fuels, but may also be required for enthalpy calculations in reacting systems.

LIQUID HEAT CAPACITY

There are basically four types of correlations for predicting liquid heat capacities: theoretical, group-contribution, corresponding-states, and empirical. In a comprehensive review of liquid heat capacity correlations, Reid et al. (1977) pointed out that reliable theoretical procedures for predicting liquid heat capacities had not yet been developed for engineering applications. In addition, although the group-contribution and corresponding-states methods have proved to be satisfactory for defined compounds, these methods cannot easily be applied to petroleum and coal-derived liquids, because of the lack of information on molecular structure and critical properties. Finally, most corresponding-states methods also require the ideal-gas heat capacity in order to predict the liquid heat capacity of a compound or mixture. Although it is known accurately for most defined compounds, the ideal-gas heat capacity of fractions must be predicted from other correlations, such as Eq. 6.1. Consequently, additional error is introduced in the final predicted liquid heat capacity value.

An example of a corresponding-states correlation for predicting the enthalpy of coal liquids is that proposed by Starling et al. (1980) (see also Brulé et al., 1982). In addition, Starling et al. proposed a procedure for predicting the ideal-gas heat capacity of coal fractions for use in their corresponding-states approach. However, Gray and Holder (1982) found that this corresponding-states correlation yielded average absolute deviations ranging from 9.1 to 39.5% for the 10 SRC-II narrow-boiling coal liquids that they were investigating.

Empirical Correlations

A simpler approach is the one proposed by Kesler and Lee (1976), who developed their method for calculating the liquid and vapor enthalpies of petroleum-derived fluids. This method included a corresponding-states correlation for predicting the liquid enthalpy in the critical region (Lee and Kesler, 1975), but uses the Watson and Nelson (1933) correlation for predicting the liquid heat capacity (and enthalpy) away from the critical region. This correlation is given by

$$C_p^L = (0.35 + 0.055\,K_w)\left[0.6811 - 0.308\,S + (0.815 - 0.306\,S)\,\frac{t}{1000}\right] \quad (6.2)$$

where C_p^L is the liquid heat capacity in British thermal units per pound per degree Fahrenheit and t is the temperature in degrees Fahrenheit. The Kesler-Lee method uses the Watson-Nelson correlation to predict the liquid enthalpy up to a reduced temperature (T/T_c) of 0.80, where T_c is the critical temperature.

A similar approach is presented in the API Technical Data Book (1983), which uses the Kesler-Lee corresponding-states correlation for vapor and near-critical region liquid enthalpies. However, a correlation different from that of Watson and Nelson is given for predicting liquid heat capacities below a reduced temperature of 0.80:

$$C_p^L = a + bT + cT^2 \tag{6.3a}$$

$$a = -1.17126 + (0.023722 + 0.024907\,S) \times K_w$$
$$+ (1.14982 - 0.046535\,K_w)/S \tag{6.3b}$$

$$b = (1 + 0.82463\,K_w)(1.12172 - 0.27634/S) \times 10^{-4} \tag{6.3c}$$

$$c = -(1 + 0.82463\,K_w)(2.90270 - 0.70958/S) \times 10^{-8} \tag{6.3d}$$

The liquid heat capacity correlation proposed by API incorporates the many petroleum fraction enthalpy data that have been reported since the development of the original Watson-Nelson correlation, and therefore should be more widely applicable.

However, neither the Watson-Nelson nor the API correlation, in its original development, included liquid heat capacity data for highly aromatic materials. Consequently, one would not expect these correlations to yield satisfactory predictions of liquid heat capacities for coal liquids. In a previous study (Exxon, 1980), we indicated that, although the Watson-Nelson correlation may not be suitable in its original form, improved predictions could be obtained with a modified correlation based on the analysis of available liquid heat capacity data for coal liquids and model compounds.

Modified Watson-Nelson Correlation for Coal Liquids

Experimental liquid heat capacity data for 28 model compounds (342 data points) and 24 coal-liquid fractions (389 data points) were used as the basis for developing a new liquid heat capacity correlation. Data for model compounds not given in Table 1.1 (e.g., benzene, toluene) were included in the analysis in order to extend the range of applicability of the final correlation. The data sources and characterization for the 28 defined compounds, along with the temperature range of the heat capacity data, are summarized in Table 6.1.

Included in this data base are liquid heat capacities obtained at Exxon for five of the coal liquid model compounds given in Table 1.1. These are acenaphthene (Mraw and O'Rourke, 1981), phenylbenzene, phenylcyclohexane, cyclohexylcyclohexane, and dibenzothiophene (O'Rourke and Mraw, 1983). Other sources of liquid heat capacity data on model compounds include the Colorado School of Mines (Kidnay and Yesavage, 1980a, 1980b, 1980c, 1981, 1982; Mohr et al., 1983) and NIPER (formerly the Bartlesville Energy Technology Center; Finke et al., 1977; Gammon et al., 1982; Lee-Bechtold et al., 1979).

These same sources also provide data for 15 of the 24 coal-liquid fractions included in the data base. Of these fifteen, the heat capacities of four coal liquids produced by the EDS process, from both Illinois and Wyoming coals, have been determined at Exxon and reported by Mraw et al. (1984). Figure 6.2 illustrates the linear temperature dependence of the heat capacity of these liquids up to about 400°F. In addition, Kidnay and Yesavage (1977a, 1977b, 1977c, 1978, 1980a, 1980b, 1980c) reported a

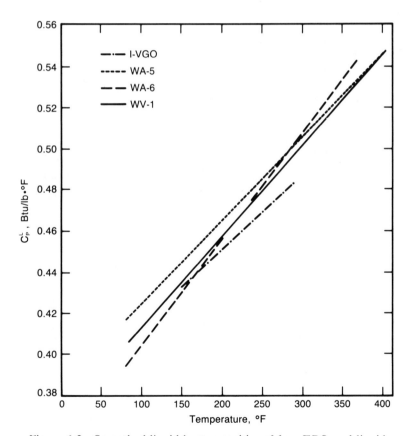

Figure 6.2 Smoothed liquid heat capacities of four EDS coal liquids.

Table 6.1 Data Sources for Liquid Heat Capacity of Coal Liquid Model Compounds

Compound	Characterization[a] (K_w/S)	Temperature Range (°F)	No. of Points	Data Sources
Benzene	9.713/0.8844	44 to 374	17	Chao (1979); San José et al. (1976); San José et al. (1976); Swanson and Chueh (1972)
Toluene	10.15/0.8718	−63 to 284	17	San José et al. (1976); Scott et al. (1962)
Ethylbenzene	10.37/0.8718	−63 to 80	9	Scott and Brickwedde (1945)
o-Xylene	10.28/0.8848	−10 to 80	6	Chao (1979)
m-Xylene	10.43/0.8687	−46 to 116	9	Pitzer and Scott (1943)
p-Xylene	10.446/0.8657	62 to 440	22	Chao (1979)
n-Propylbenzene	10.62/0.8665	−63 to 206	16	Messerly et al. (1965)
Isopropylbenzene	10.57/0.8663	80 to 200	7	Schlinger and Sage (1952)
1,2,3-Trimethylbenzene	10.37/0.8987	−10 to 80	6	Taylor et al. (1955)
1,2,4-Trimethylbenzene	10.54/0.8803	−46 to 220	10	Helfrey et al. (1955); Putnam and Kilpatrick (1957)
1,3,5-Trimethylbenzene	10.63/0.8696	−46 to 80	8	Chao (1979)
Naphthalene	9.32/1.030	177 to 206	3	McCullough et al. (1957)
1-Methylnaphthalene	9.536/1.0244	−10 to 170	11	McCullough et al. (1957)
Tetralin	9.785/0.9739	−28 to 116	9	McCullough et al. (1957)
cis-Decalin	10.488/0.9011	−28 to 170	12	McCullough et al. (1957)
trans-Decalin	10.747/0.874	−10 to 170	11	McCullough et al. (1957)

	K_w/S	Range		Reference
Phenylbenzene	9.57/1.027	152 to 404	15[b]	O'Rourke and Mraw (1983)
Phenylcyclohexane	10.27/0.947	44 to 386	20[b]	O'Rourke and Mraw (1983)
Cyclohexylcyclohexane	10.92/0.889	44 to 386	20[b]	O'Rourke and Mraw (1983)
Acenaphthene	9.11/1.094	200 to 332	9	Finke et al. (1977)
Phenanthrene	9.15/1.130	211 to 296	6	Finke et al. (1977)
9,10-Dihydrophenanthrene	9.35/1.09	92 to 170	6	Lee-Bechtold et al. (1979)
Anthracene	9.21/1.123	434 to 461	4	Goursot et al. (1970)
1,2,3,4,5,6,7,8-Octahydroanthracene	9.87/1.02	162 to 260	7	Gammon et al. (1982)
Phenol[d]	8.647/1.082	105 to 134	3	Andon et al. (1963)
o-Cresol[d]	8.961/1.051	88 to 260	11	Andon et al. (1967)
m-Cresol[d]	9.142/1.0385	100 to 600	26[c]	Mohr et al. (1983)
2,6-Dimethylpyridine	9.848/0.923	100 to 500	21[c]	Kidnay and Yesavage (1981); Mohr et al. (1983)
Quinoline	8.851/1.0986	100 to 700	31[c]	Kidnay and Yesavage (1982)
Thiophene	8.052/1.072	200 to 400	11[c]	Kidnay and Yesavage (1980c)
Dibenzothiophene	8.65/1.19	206 to 530	19[b]	O'Rourke and Mraw (1983)

[a]K_w is Watson characterization factor (Eq. 1.1) and S is the specific gravity at 60/60°F; see Tables 1.1 and 3.4.

[b]Smoothed heat capacity data.

[c]Smoothed data from experimental liquid enthalpies.

[d]Not included in correlation development.

Table 6.2 Data Sources for Liquid Heat Capacity of Coal Liquids

Coal Liquid	Characterization[a] (K_w/S)	Temperature Range (°F)	No. of Points	Data Sources
EDS 1-VGO	9.82/1.118	160–300	8[b]	Mraw et al. (1984)
EDS WA-5	10.14/0.970	80–400	17[b]	Mraw et al. (1984)
EDS WA-6	10.02/0.987	80–360	15[b]	Mraw et al. (1984)
EDS WV-1	9.96/0.998	80–400	17[b]	Mraw et al. (1984)
COED Western Kentucky	10.90/0.923	100–700	31[c]	Kidnay and Yesavage (1977a)
COED Western Kentucky distillate	10.60/0.884	100–500	21[c]	Kidnay and Yesavage (1977b)
COED Utah distillate	10.68/0.879	100–500	21[c]	Kidnay and Yesavage (1977b)
COED Western Kentucky syncrude				
<401°F distillate	10.99/0.823	60–300	13[b]	Smith et al. (1980)
401–716°F distillate	10.56/0.941	60–460	21[b]	
COED Utah A-seam syncrude				
<399°F distillate	11.03/0.824	60–200	8[b]	Smith et al. (1980)
399–718°F distillate	10.67/0.938	60–460	21[b]	

SRC-I Synthoil distillate	10.00/0.978	100–600	26[c]	Kidnay and Yesavage (1977c)
SRC-I naphtha	11.30/0.781	100–400	16[c]	Kidnay and Yesavage (1978)
SRC-II naphtha	10.75/0.820	100–700	31[c]	Kidnay and Yesavage (1980b)
SRC-II middle distillate	9.90/0.976	100–600	2[c]	Kidnay and Yesavage (1980b)
SRC-II Narrow Boiling Coal Liquids				Gray and Holder (1982)
Cut 2	11.37/0.770	77–302	6	
Cut 4	11.13/0.813	77–392	8	
Cut 6	9.91/0.954	77–392	8[d]	
Cut 8	10.05/0.976	77–482	10	
Cut 10	10.13/0.997	77–527	11	
Cut 12	9.77/1.079	77–617	13	
Cut 14	9.92/1.018	77–572	12	
Cut 15	9.74/1.077	77–617	13	
Cut 17	9.70/1.120	77–617	13	
Cut 19	9.43/1.195	77–572	11	

[a]See Table 6.1 for definition.

[b]Smoothed heat capacity data.

[c]Smoothed data from experimental liquid enthalpies.

[d]Not included in correlation development.

substantial body of enthalpy data on coal liquids produced by the COED, SRC-I, and SRC-II processes, and Smith et al. (1980) determined liquid heat capacities for several COED liquid products. Finally, data for the remaining nine coal liquids were reported by Gray and Holder (1982), who measured liquid heat capacities for 10 narrow-boiling liquids obtained from SRC-II. (The tenth was excluded because it contains a high concentration of phenolic compounds, but is included in the comparison of correlation predictions with experimental data.) The data sources, characterization, and temperature range of the heat capacity data for the 24 coal liquid fractions are summarized in Table 6.2.

Our analysis of these data led to the following modified Watson-Nelson correlation, shown in Figure 6.3:

$$C_p^L = (0.06759 + 0.05638 K_w)$$

$$\times \left[0.6450 - 0.05959 S + (1.2892 - 0.5264 S) \frac{t}{1000} \right] \quad (6.4)$$

The correlation accuracy for model compounds, coal liquids, and the total data base is summarized in Table 6.3. Note that the original Watson-Nelson correlation, Eq. 6.2, reproduces the liquid heat capacity data surprisingly well both for model compounds and overall. The AAD (average absolute deviation) of 4.52% and 6.40% for the model compounds

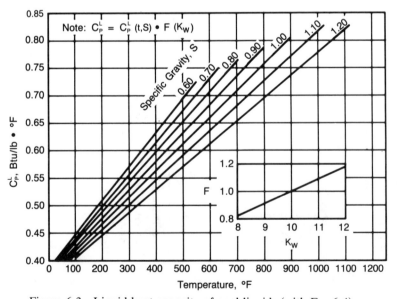

Figure 6.3   Liquid heat capacity of coal liquids (with Eq. 6.4).

Table 6.3 Summary of Deviations in Predicted Liquid Heat Capacities

	% Deviation[a]								
	Model Compounds			Coal Liquids			Total		
	AAD	Bias	Max	AAD	Bias	Max	AAD	Bias	Max
Original Watson-Nelson (Eq. 6.2)	4.52	−2.49	−12.10	8.05	−7.79	−21.91	6.40	−5.31	−21.91
API (Eq. 6.3)	6.16	−4.16	−20.12	8.34	−7.99	−23.90	7.32	−6.20	−23.90
Modified Watson-Nelson (Eq. 6.4)	4.13	3.13	12.82	3.66	−1.66	−16.29	3.88	0.58	−16.29
Number of data points	342			389			731		

[a] % Deviation $= 100 \times (\text{calc} - \text{exp})/\text{exp}$; AAD $= [\sum_i |\% \text{ Dev}_i|]/N$; Bias $= [\sum_i \% \text{ Dev}_i]/N$; N = number of data points.

Table 6.4 Comparison of Watson–Nelson Correlations with Experimental Liquid Heat Capacity for Coal Liquid Model Compounds

Compound	No. of Points	Original, Equation 6.2			Modified, Equation 6.4		
		AAD	Bias	Max	AAD	Bias	Max
Benzene	17	3.05	-3.05	-5.72	1.73	-1.73	-4.41
Toluene	17	2.10	0.54	-5.04	3.15	1.95	4.90
Ethylbenzene	9	2.17	-2.13	-5.54	1.91	-0.61	-4.59
o-Xylene	6	3.72	-3.72	-5.45	1.55	-1.47	-3.49
m-Xylene	9	1.75	0.74	-3.08	2.89	2.42	5.22
p-Xylene	22	4.59	4.59	5.10	7.49	7.49	8.18
n-Propylbenzene	16	1.84	-0.72	-5.82	2.91	1.58	5.01
Isopropylbenzene	7	2.18	2.18	2.85	4.82	4.82	5.15
1,2,3-Trimethylbenzene	6	5.92	-5.92	-7.74	2.65	-2.65	-4.71
1,2,4-Trimethylbenzene	10	3.47	-2.04	-6.67	3.15	0.76	6.02
1,3,5-Trimethylbenzene	8	1.39	0.90	2.73	3.10	3.10	5.41
Naphthalene	3	3.59	-3.59	-3.75	4.80	4.80	4.90
1-Methylnaphthalene	11	6.14	-6.14	-8.27	2.80	2.80	3.77
Tetralin	9	4.98	-4.98	-5.67	1.61	1.61	1.80

% Deviation[a]

cis-Decalin	12	2.98	2.98	3.95	7.18	7.18	7.95
trans-Decalin	11	8.27	8.27	9.90	11.69	11.69	12.82
Phenylbenzene	15	4.24	−4.24	−6.74	4.02	4.02	5.79
Phenylcyclohexane	20	0.72	−0.65	−1.45	5.72	5.72	6.89
Cyclohexylcyclohexane	20	2.26	1.91	6.01	7.14	7.14	10.79
Acenaphthene	9	7.47	−7.47	−7.96	3.11	3.11	3.19
Phenanthrene	6	7.62	−7.62	−8.05	5.01	5.01	5.09
9,10-Dihydrophenanthrene	6	10.05	−10.05	−10.90	1.87	1.87	2.25
Anthracene	4	5.25	−5.25	−5.45	5.51	5.51	5.61
1,2,3,4,5,6,7,8-Octahydroanthracene	7	6.15	−6.15	−6.27	2.60	2.60	3.01
Phenol[b]	3	35.13	−35.13	−35.41	28.56	−28.56	−28.76
o-Cresol[b]	11	30.69	−30.69	−34.38	24.48	−24.48	−28.00
m-Cresol[b]	26	23.72	−23.72	−34.28	17.82	−17.82	−28.11
2,6-Dimethylpyridine	21	4.71	−4.71	−8.45	2.01	−0.96	−5.07
Quinoline	31	8.45	−8.45	−12.10	0.57	0.11	−1.16
Thiophene	11	5.28	−4.74	−10.53	4.06	0.43	8.70
Dibenzothiophene	19	3.80	−3.80	−7.20	10.23	10.23	11.44

[a]See Table 6.3.
[b]Not included in development of Eq. 6.4.

Table 6.5 Comparison of Watson-Nelson Correlations with Experimental Liquid Heat Capacity for Coal Liquids

Coal Liquid	No. of Points	Original, Equation 6.2			Modified, Equation 6.4		
		AAD	Bias	Max	AAD	Bias	Max
EDS 1-VGO	8	15.04	−15.04	−16.80	2.56	−2.56	−3.78
EDS WA-5	17	4.23	−4.23	−7.44	2.74	2.69	5.34
EDS WA-6	15	4.69	−4.69	−5.41	2.80	2.80	4.20
EDS WV-1	17	5.98	−5.98	−7.62	1.81	1.81	2.76
COED Western Kentucky	31	3.12	−1.24	−5.79	5.64	5.64	12.69
COED Western Kentucky distillate	21	3.50	−3.50	−3.62	0.57	0.57	1.00
COED Utah distillate	21	6.77	−6.77	−11.98	3.51	−2.84	−8.97
COED Western Kentucky syncrude							
<401°F distillate	13	2.90	−2.90	−4.75	1.68	−1.26	−4.12
401–716°F distillate	21	5.03	−5.03	−7.14	1.63	1.49	3.29
COED Utah A-seam syncrude							
<399°F distillate	8	5.04	−5.04	−6.49	3.65	−3.65	−4.58
399–718°F distillate	21	9.18	−9.18	−10.89	2.82	−2.82	−4.71

% Deviation[a]

SRC-I Synthoil distillate	26	13.78	−13.78	−20.38	7.69	7.69	−14.30
SRC-I naphtha	16	3.67	−3.67	−6.67	2.91	2.91	−7.06
SRC-II naphtha	31	6.83	−6.83	−13.19	5.01	5.01	−12.94
SRC-II middle distillate	26	13.13	−13.13	−21.91	7.36	7.36	−16.29
SRC-II Narrow Boiling Coal Liquids							
Cut 2	6	3.76	3.76	4.73	3.93	3.93	4.56
Cut 4	8	3.96	−3.96	−6.05	2.15	−2.11	−5.50
Cut 6[b]	8	17.23	−17.23	−19.10	12.24	−12.24	−14.12
Cut 8	10	13.47	−13.47	−16.60	7.19	−7.19	−10.14
Cut 10	11	9.95	−9.95	−12.56	2.21	−2.21	−5.62
Cut 12	13	11.57	−11.57	−15.43	1.50	−1.09	−5.24
Cut 14	12	10.91	−10.91	−13.23	2.88	−2.88	−4.92
Cut 15	13	12.20	−12.20	−17.37	2.95	−1.99	−5.60
Cut 17	13	14.14	−14.14	−16.96	2.77	−2.08	−5.87
Cut 19	11	17.05	−17.05	−19.11	2.25	−2.13	−4.26

[a]See Table 6.3.

[b]Not included in development of Eq. 6.4.

and total data base, respectively, suggests that the Watson-Nelson correlation can be applied to coal liquids with reasonable accuracy. However, the relatively large negative bias (-7.79% for coal liquids) indicates that the original correlation generally underpredicts their liquid heat capacity.

Also shown in Table 6.3 are the deviations for the API liquid heat capacity correlation, Eq. 6.3, and the modified Watson-Nelson correlation, Eq. 6.4. The accuracy of the API correlation is slightly lower than that of the original Watson-Nelson correlation (7.32 versus 6.40% AAD), and the API correlation exhibits a slightly larger negative bias (-6.20 versus -5.31%). Much better results are obtained from the modified form of the Watson-Nelson correlation, which yields average deviations of about 3.9% and substantially reduces the overall bias to about $+0.58\%$. There is also substantial improvement in the coal liquid heat capacities predicted by Eq. 6.4 versus 6.2 (3.66 versus 8.05% AAD).

We also investigated several modifications to the API correlation, both including and excluding the quadratic temperature term. However, we found that these modifications produced only minor improvements in accuracy over the modified Watson-Nelson form. In addition, we observed that the addition of a quadratic temperature term did not significantly improve the accuracy of the predicted liquid heat capacities below a reduced temperature of 0.80.

Tables 6.4 and 6.5 present summaries of deviations by compound for the model compounds and coal liquids, respectively. Note that the liquid heat capacities of the 10 SRC-II narrow-boiling coal liquids (Gray and Holder, 1982) are reasonably well predicted by both the original Watson-Nelson correlation and Eq. 6.4 (average deviations range from 3.76 to 17.23% and 1.50 to 12.24%, respectively). These compare very favorably to the 9.1 to 39.5% range of average deviations obtained from the correlation of Starling et al. (1980), as previously discussed.

Table 6.5 also shows that the experimental liquid heat capacities reported by Mraw et al. (1984) for the EDS-produced Illinois and Wyoming coal liquids are predicted very accurately by Eq. 6.4. Deviations for these four liquids range from 1.81 to 2.80% with a maximum error of 5.34%. Considering the experimental uncertainty of the reported measurements, these results suggest that the coal source has little effect on the accuracy of the liquid heat capacity predictions. Similarly, Eq. 6.4 reproduces nearly all of the coal liquid heat capacities from other processes (COED, SRC-I, SRC-II) within experimental uncertainty. This indicates that, aside from the effects due to boiling point and gravity, the heat capacity of coal liquids is not strongly dependent upon the liquefaction process used to produce these liquids.

In addition, with the exception of phenolic compounds, the heat capacity of coal liquids does not appear to be strongly dependent upon their heteroatom content. As discussed in Chapter 1, a significant difference between coal liquids and petroleum-derived liquids is the level of hetero-atoms. Coal liquids generally contain much less sulfur, but more nitrogen, and substantially more oxygen (principally in phenolic compounds) than petroleum liquids do. However, our analysis indicates that the two-parameter (S and K_w) characterization results in an accurate correlation of liquid heat capacity data for most coal liquids except those containing high levels of phenolic compounds.

To illustrate this point, Table 6.4 includes deviations for phenolic com-pounds (phenol, cresols) not included in our correlation development. Phenolic compounds have been shown (Gray and Holder, 1982) to be present in significant quantities in SRC-II coal liquids boiling from about 350 to 450°F (note that phenol boils at about 359°F and cresols at 376–396°F). These compounds are highly associated and, as shown in Table 6.4, their liquid heat capacities are poorly predicted by all of the correlations evaluated. Also note the relatively large deviations for the SRC-II narrow-boiling Cut 6 in Table 6.5. Cut 6 boils at a nominal 384°F and contains a large proportion of phenolic compounds, thus accounting for the unusually large deviations in the predicted liquid heat capacity.

These observations clearly indicate that the two-parameter charac-terization is insufficient for predicting the liquid heat capacity of these highly associated compounds. A third parameter is necessary that will reflect the polar nature of the phenol-containing coal-liquid fractions. Sharma (1980) suggested that the cryoscopic molecular weight deter-mination (i.e., the freezing point depression method of measuring molecular weight) as a function of benzene concentration might be one means of characterizing association. Furthermore, Sharma showed that the slope of the molecular weight–benzene concentration curve from this method was directly related to errors in the prediction of SRC-II enthalpies with the Kesler and Lee (1976) procedure previously discussed. Clearly, more work is needed to properly characterize association effects in coal liquids. This is also noted in Chapter 10.

HEAT OF VAPORIZATION

Once the liquid and ideal-gas heat capacities are known, one needs to establish the difference between the liquid and ideal-gas enthalpies to calculate the heat of vaporization. In the method developed by Kesler and Lee (1976), the Watson-Nelson correlation is used to predict the liquid heat capacity (and enthalpy) up to a reduced temperature (T_r) of 0.80. At higher reduced temperatures, the liquid enthalpy is predicted by the correlation developed by Lee and Kesler (1975)—a three-parameter (T_c, P_c, ω) corresponding states correlation which is also used for all vapor-phase enthalpy calculations. The Lee-Kesler correlation predicts the departures of enthalpy and heat capacity from the ideal-gas behavior as a function of reduced temperature, reduced pressure (P_r), and acentric factor (ω):

$$\frac{H^* - H}{RT_c} = f'(T_r, P_r, \omega)$$

$$\frac{C_p - C_p^*}{R} = f''(T_r, P_r, \omega)$$

Consequently, it can also be used to predict the heat of vaporization.

However, since Eq. 6.4 is to be used to predict the liquid enthalpy and heat capacity below reduced temperatures of 0.80 and, furthermore, it must be continuous with the Lee-Kesler correlation at a reduced temperature of 0.80, the ideal-gas enthalpy must be established from the Lee-Kesler correlation enthalpy departure at $T_r = 0.80$ and at saturation pressure. This is illustrated in Figure 6.4, which shows the liquid enthalpy given by Eq. 6.4 from $T_r = 0$ up to $T_r = 0.80$. At $T_r = 0.80$, the Lee-Kesler predicted enthalpy departure at saturation pressure is given by

$$\left(\frac{H^* - H^L}{RT_c}\right)_{T_r=0.8} = 4.5092 + 5.264\omega \tag{6.5}$$

where H^* is the ideal-gas enthalpy, H^L is the liquid enthalpy, T_c is the critical temperature, and ω is the acentric factor. The ideal-gas enthalpy is obtained by integrating Eq. 6.1 with respect to temperature, and by establishing the value of the integration constant from the value of H^* at $T_r = 0.80$ as given by

$$H^* = H^L + RT_c(4.5092 + 5.264\omega) \tag{6.6}$$

where H^L is given by Eq. 6.4 at $T_r = 0.80$. With the ideal-gas enthalpy basis established, the heat of vaporization is given by

$$\Delta H_V = H^* - (H^* - H^V)_{LK} - H^L, \quad T_r \leq 0.80 \tag{6.7}$$

where $(H^* - H^V)_{LK}$ is the Lee-Kesler-predicted enthalpy departure for the saturated vapor and H^L is the liquid enthalpy obtained from the modified Watson-Nelson correlation, Eq. 6.4. At temperatures above $T_r = 0.80$, the heat of vaporization is given by the Lee-Kesler correlation

$$\Delta H_V = (H^* - H^L)_{LK} - (H^* - H^V)_{LK} \tag{6.8}$$

where $(H^* - H^L)_{LK}$ is the predicted enthalpy departure for the saturated liquid.

Since the value of $(H^* - H^V)_{LK}$ is typically less than 5% of ΔH_V for $T_r < 0.80$, Eq. 6.7 shows that the accuracy of the predicted heat of

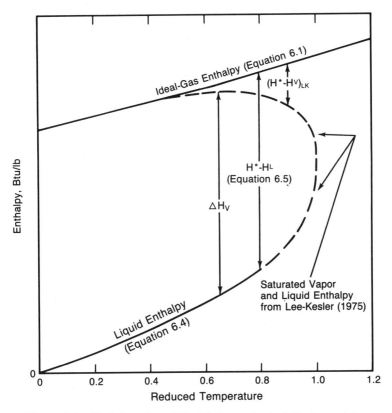

Figure 6.4   Enthalpy of coal liquids by method of Kesler and Lee.

vaporization as a function of temperature depends largely on the accuracy of $H^* - H^L$. Furthermore, since Eqs. 6.1 and 6.4 were shown to accurately represent the ideal gas and liquid heat capacities of coal liquid materials, the accuracy of $H^* - H^L$ is determined by Eq. 6.5, the Lee-Kesler predicted enthalpy departure of the saturated liquid at $T_r = 0.80$. Although values of $(H^* - H^L)$ are not generally available for most materials, we can infer the accuracy of Eq. 6.5 by comparing experimental values of the heat of vaporization of coal liquids and model compounds at $T_r = 0.80$ with those values predicted by the Lee-Kesler correlation, which gives

$$(\Delta H_V)_{T_r = 0.80} = RT_c(4.0439 + 5.3826\omega) \qquad (6.9)$$

Table 6.6 presents experimental heats of vaporization at $T_r = 0.80$ for several model compounds and coal liquid fractions, and compares these with values obtained from Eq. 6.9. The critical temperatures and acentric factors were obtained from the recommended procedures given in Chapters

Table 6.6   Experimental and Predicted Heats of Vaporization for Model
Compounds and Coal Liquids

	Heat of Vaporization at $T_r = 0.80$ (in Btu/lb)			
	Exp	Calc	% Error	Data Sources
Benzene	136.0	133.6	−1.76	Kidnay and Yesavage (1980a)
m-Xylene	123.0	120.6	−1.99	Kidnay and Yesavage (1980a)
n-Butylbenzene	108.4	109.6	−1.13	Svoboda et al. (1981)
Naphthalene	113.4	118.4	4.41	Kudchadker et al. (1978b)
Tetralin	111.4	109.9	−1.31	Kudchadker et al. (1978a)
1-Methylnaphthalene	119.0	113.2	−4.84	Kidnay and Yesavage (1980a)
Diphenylmethane	105.9	104.8	−1.04	Wieczorek and Kobayashi (1981)
Phenol	162.6	169.1	4.00	Kudchadker et al. (1977)
m-Cresol	154.3	150.9	−2.20	Andon et al. (1960); Mohr et al. (1983)
2,4-Xylenol	129.5	141.0	8.88	Kudchadker and Kudchadker (1978)
Diphenyl ether	102.9	105.5	2.53	Grebe (1932)
Quinoline	131.2	126.0	−3.96	Kidnay and Yesavage (1982)
SRC-I naphtha	117.0	116.3	−0.62	Kidnay and Yesavage (1980a)
SRC-II				
5HC	108.5	115.0	6.01	Gray et al. (1983)
8HC	106.2	106.0	−0.17	Gray et al. (1983)
11HC	103.3	97.4	−5.73	Gray et al. (1983)
16HC	88.6	95.2	7.47	Gray et al. (1983)
17HC	91.8	93.4	1.76	Gray et al. (1983)
4HC-A	104.4	113.2	8.45	Gray and Holder (1982)
6HC	111.8	114.3	2.24	Gray and Holder (1982)
7HC-B	104.4	110.7	6.03	Gray and Holder (1982)
10HC-B	90.4	100.3	10.91	Gray and Holder (1982)
15HC-B	92.2	98.1	6.40	Gray and Holder (1982)
Total (23 points)			4.08	

3 and 4. These comparisons show that errors in the heats of vaporization predicted by Eq. 6.9 are typically less than 5%, suggesting that the enthalpy departure given by Eq. 6.5 is of acceptable accuracy. Therefore, the heats of vaporization predicted by Eqs. 6.7 (for $T_r \leq 0.8$) and 6.8 (for $T_r > 0.8$) should be of similar accuracy.

Included in Table 6.6 are comparisons of predicted and "experimental" heats of vaporization for selected phenolic and other oxygenated compounds. Although most "experimental" ΔH_V values at $T_r = 0.8$ were estimated from lower temperature data (generally, below the compound's normal boiling point), the values predicted by Eq. 6.9 are in surprisingly good agreement. Furthermore, at $T_r > 0.8$, comparisons of ΔH_V predicted by Eq. 6.8 for highly associating C_2–C_4 alcohols, for which experimental data are available, show good agreement with the experimental data (although similar comparisons with Eq. 6.8 at $T_r < 0.8$ show relatively poor agreement with data). These comparisons illustrate our need to better understand and characterize these association effects in coal liquids, especially those containing significant quantities of phenolic compounds.

HEAT OF COMBUSTION

The heat of combustion of coal liquids is an important property for two reasons. First, the heat of combustion is used to determine the quality of the hydrocarbon liquid as a fuel source. The "heating value" of the liquid is given by the gross heat of combustion (i.e., the product water is liquid) and ranges from 15 to 20 kBtu/lb. Heating values are typically predicted with uncertainties of less than ±2%. The heat of combustion value can also be used to derive a heat of formation for reaction enthalpy calculations. This second application is much more sensitive to the accuracy of the predicted heat of combustion and may require maximum uncertainties of less than ±0.5%. This will be discussed later in this section.

Two types of correlations are generally available for predicting the heat of combustion of coal liquids. The first, which is less frequently used, expresses the heat of combustion as a function of specific gravity or °API and the Watson characterization factor, K_w. A graphical correlation of this type is given by Hougen, Watson, and Ragatz (1954) for petroleum-derived liquids. The API Technical Data Book (1983) provides an analytical correlation, developed in 1960, that includes specific gravity, K_w, wt% hydrogen, and corrections for heteroatoms, water, and mineral (ash) material. Both of these correlations predict the heat of combustion of petroleum liquids with an average uncertainty of about ±200 Btu/lb. This uncertainty is generally acceptable for heating value calculations, but is unacceptable for heat of reaction calculations.

More recent developments in correlating and predicting heats of combustion of liquid (and solid) fuels have focused on the correlation model originally proposed by Dulong, as discussed by Francis and Lloyd (1983):

$$-\Delta H_C = a_0 + a_1(C) + a_2(H) + a_3(S) + a_4(N) - a_5(O) \qquad (6.10)$$

where ΔH_C is the gross heat of combustion (kBtu/lb), C, H, S, N, and O are the weight percent of these elements in the (organic) material, and a_0–a_5 are constants derived from experimental values of heats of combustion. In Dulong's original formulation, $a_0 = a_4 = 0$. Mott and Spooner (1940) used Eq. 6.10 and developed a set of constants for predicting the heating values of coals.

One of the more successful empirical correlations is that of Boie, as given by Ringen et al. (1979), who developed the following equation for application to both solid and liquid fuels:

$$-\Delta H_C = 0.1512(C) + 0.4998(H) + 0.045(S) + 0.027(N) - 0.0447(O) \quad (6.11)$$

Ringen et al. (1979) compared predictions from Eq. 6.11 with heat of combustion data for a wide variety of fuels, including naphtha and distillate fuels, coals, shale oils, and residua, and obtained average errors of about $\pm 1.4\%$ for the liquid fuels. The largest errors were for the solid fuels and averaged about 1.9%. Considering the general applicability of the Boie correlation, these results are very encouraging and suggest that more accurate predictions can be obtained by limiting the correlation development to either liquid or solid fuels.

Two recently proposed correlations appear to provide significantly improved predictions for the heat of combustion of liquid fuels, including coal-derived liquids and model compounds. Lloyd and Davenport (1980) developed their correlation from heat of combustion data for 138 compounds representative of those found in fossil fuel liquids and, particularly, coal-derived liquids. Furthermore, their data base included a substantial number of heteroatom-containing compounds of all types (S, N, and O). Their correlation is given by

$$-\Delta H_C = 0.1538(C) + 0.4883(H) + 0.04811(S) + 0.02554(N) - 0.03633(O) \qquad (6.12)$$

A correlation similar to Eq. 6.12 was developed by Jain and Sundararajan (1981), who observed that heats of reaction of stoichiometric fuel-oxidizer systems were linearly related to the total oxidizing or reducing valencies of the mixture. Furthermore, they demonstrated that the gross heat of combustion was directly related to the net reducing valencies of the fuel. Although the details are omitted here, their heat of combustion correlation can be expressed by the following equation:

$$-\Delta H_C = 0.1408(C) + 0.4224(H) + 0.05279(S) - 0.05279(O) + 1.8426$$

$$(6.13)$$

Predictions from Eqs. 6.12 and 6.13 have been compared against a data base of 130 experimental heats of combustion at the standard reference state of 77°F (25°C) for coal liquid model compounds and coal-derived liquid products. These include data for several of the condensed aromatic and heteroatom-containing model compounds listed in Table 1.1, data reported by Smith et al. (1980) on products from several different liquefaction processes, as well as a substantial data base for products obtained from the EDS liquefaction process (Exxon, 1978, 1981a, 1981b, 1982).

These comparisons are summarized in Table 6.7 and show that both Eqs. 6.12 and 6.13 generally underpredict the heat of combustion of coal liquids by nearly 200 Btu/lb. In the case of Eq. 6.12, this bias appears to be the result of correlating ideal-gas heats of combustion rather than values for the corresponding liquids. Consequently, Eq. 6.12 predicts heats of combustion that are generally low by quantities equivalent to the compound's heat of vaporization (150–250 Btu/lb). Unfortunately, there is no such obvious explanation for the bias of Eq. 6.13, as Jain and Sundararajan appear to have developed their correlation from heats of combustion for liquid fuels. Nevertheless, Table 6.7 shows that Eq. 6.13 is inferior to Eq. 6.12 in predicting heats of combustion for coal liquids.

However, neither correlation appears suitable for the accurate heat of

Table 6.7   Comparison of Heat of Combustion Correlations for Coal Liquids[a,b]

	% Average Deviation	% Bias	Average Error (Btu/lb)
Equation 6.12 (Lloyd and Davenport)	1.46	−1.04	210
Equation 6.13 (Jain and Sundararajan)	1.79	−1.56	230
Equation 6.14	0.55	0.01	94

[a] See Table 6.3.

[b] Data base obtained from the following sources for 30 coal liquid model compounds: American Petroleum Institute (1983), Cox and Pilcher (1970), Domalski (1982); and for 100 coal liquid fractions: Bowden and Brinkman (1980), Comilli et al. (1979), Exxon (1978, 1981a,b, 1982), Smith et al. (1980).

combustion predictions required for estimating the heat of formation, as is discussed below. Based on the same 130 heat-of-combustion data for various liquefaction products and model compounds, the following correlation was developed for predicting accurate heats of combustion for coal liquids

$$-\Delta H_C = 0.15075(C) + 0.4924(H) + 0.08829(S) + 0.02654(N) - 0.03752(O)$$
$$(6.14)$$

Table 6.7 shows that Eq. 6.14 predicts heats of combustion with average errors of about 94 Btu/lb and no appreciable bias. This represents better than a twofold improvement in accuracy when compared to Eqs. 6.12 and 6.13. Such an improvement is *essential* for the calculation of heats of formation.

Heat of Formation for Reaction Enthalpy Calculations

The main difficulty in calculating the heat of reaction for conversion processes involving coal liquids is to estimate accurately the heats of formation of the "undefined" liquid fractions. These heats of formation can be obtained from heats of combustion and elemental analyses for the specific fractions according to the exact relationship:

$$-\Delta H_f^o = 0.140848(C) + 0.609586(H) + 0.080722(S) + \Delta H_C - (H^* - H^L)$$
$$(6.15)$$

where ΔH_f^o is the ideal-gas heat or enthalpy of formation, and $(H^* - H^L)$ is the heat required to vaporize the liquid to the *ideal* gas state. This calculation is typically performed at 77°F. However, for performing the reaction enthalpy calculations at reaction conditions, sensible and latent heat effects are also required.

Inspection of Eqs. 6.14 and 6.15, as well as of standard tables of heats of formation and combustion for pure compounds (e.g., Stull et al., 1969), indicates that the heat of formation typically is a small fraction of the heat of combustion. Consequently, small errors in the heat of combustion (and the elemental analysis) of a coal liquid fraction will translate, via Eq. 6.15, into very large errors in the calculated heat of formation. Therefore, very accurate elemental analyses and heats of combustion are required in order to obtain acceptable values of heats of formation. The maximum allowable uncertainty will depend upon the magnitude of the heat of reaction, but errors in these variables should generally be less than ±0.5% for reliable heat of reaction calculations. Equation 6.14 approaches this limit and can further be improved by obtaining additional experimental heat of combustion data for selected coal liquid fractions.

An alternative approach to heat of reaction calculations, as applied to the EDS process, has been discussed by Joswick and Madden (1983). This approach is based on selected hydrogenation and dehydrogenation reactions for aromatic model compounds and estimates the heat of reaction from the amount of hydrogen consumed per unit of feed to the reactor. Although this procedure appears to work well for the hydroconversion reactions that take place in the EDS process, it cannot easily be generalized for other types of reactions.

Finally, in addition to uncertainties in heats of combustion or formation, errors in sensible and latent heat effects may also diminish the reliability of heat of reaction calculations. However, the effect of these errors is much less severe than those arising from the uncertainties in the thermochemical data. Although the maximum allowable uncertainty will, again, depend upon the magnitude of the heat of reaction, errors of ±2 and 5% in heat capacity and heat of vaporization, respectively, are generally acceptable for reasonably accurate heat of reaction predictions.

CONCLUSIONS AND RECOMMENDATIONS

For ideal-gas heat capacity and enthalpy, Eq. 6.1 should provide reliable estimates over the temperature range of interest for coal liquid systems. Although the level of polynuclear aromatic and associating (e.g., phenolic) compounds in the vapor phase is generally very low, Eq. 6.1 was shown to predict reasonably accurate heat capacities for model compounds representative of these compound types. Therefore, the ideal-gas heat capacity and enthalpy should be well predicted by Eq. 6.1.

Our analysis of experimental heat capacity data for defined model compounds and coal-derived liquids resulted in a modified Watson-Nelson correlation, Eq. 6.4, for liquid heat capacity. This correlation reduces the average deviation in predicted heat capacities of coal liquids from 8.05% for the original Watson-Nelson correlation, Eq. 6.2, to 3.66%. This correlation should be used in place of the original Watson-Nelson correlation in the Kesler and Lee (1976) method for calculating the liquid enthalpy of coal liquids. Additional work is needed to properly account for the effect of association on the liquid heat capacity of coal liquids that contain significant quantities of phenolic compounds.

When used with Eq. 6.4 for the liquid heat capacity of coal-liquid fractions, the method of Kesler and Lee (1976) can be expected to predict heats of vaporization with average errors of about ±5%. Larger errors may occur for highly associating fractions such as those containing large quantities of phenols.

Finally, heats of combustion and heats of formation should be obtained from Eqs. 6.14 and 6.15, respectively. Average errors of about 100 Btu/lb should be expected. Although this accuracy is most likely sufficient for heating value calculations, it may be unacceptable for heat of reaction calculations. Consequently, further improvements in heat of reaction calculations may require experimental combustion data for selected fractions.

REFERENCES

American Petroleum Institute, *Technical Data Book—Petroleum Refining*, 4th Ed., API, Washington, DC, 1983, Chap. 7.

Andon, R. J. L., D. P. Biddiscombe, J. D. Cox, R. Handley, D. Harrop, E. F. G. Herington, and J. F. Martin, Thermodynamic Properties of Organic Oxygen Compounds. Part 1. Preparation and Physical Properties of Pure Phenol, Cresols, and Xylenols, *J. Chem. Soc.*, 5246 (1960).

Andon, R. J. L., J. F. Counsell, E. F. G. Herington, and J. F. Martin, Thermodynamic Properties of Organic Oxygen Compounds. Part 7. Calorimetric Study of Phenol from 12 to 330 K, *Trans. Faraday Soc.*, **59**, 830 (1963).

Andon, R. J. L., J. F. Counsell, E. B. Lees, J. F. Martin, and C. J. Mash, Thermodynamic Properties of Organic Oxygen Compounds. Part 17. Low-Temperature Heat Capacity and Entropy of the Cresols, *Trans. Faraday Soc.*, **63**, 1115 (1967).

Bowden, J. N., and D. W. Brinkman, Stability of Alternate Fuels, *Hydrocarbon Process.*, **59**(7), 77 (1980).

Brulé, M. R., C. T. Lin, L. L. Lee, and K. E. Starling, Multiparameter Corresponding-States Correlation of Coal-Fluid Thermodynamic Properties, *AIChE J.*, **28**, 616 (1982).

Chao, J., Properties of Alkylbenzenes, *Hydrocarbon Process.*, **58**(11), 295 (1979).

Comilli, A. G., E. S. Johanson, and P. Sheth, H-Coal PDU Tests of Illinois #6 and Wyoming Coals, EPRI Report AF-1143-SR, June 1979.

Cox, J. D., and G. Pilcher, *Thermochemistry of Organic and Organometallic Compounds*, Academic Press, New York, 1970.

Domalski, E. S., Selected Values of Heats of Combustion and Heats of Formation of Organic Compounds Containing the Elements C, H, N, O, P, and S, *J. Phys. Chem. Ref. Data*, **1**, 221 (1972).

Exxon Research and Engineering Company, EDS Coal Liquefaction Process Development. Phase IIIA. Final Technical Progress Report for the Period January 1, 1976–June 30, 1977, FE-2353-20, February 1978.

———, Phase IV. Quarterly Technical Progress Report for the Period October–December 1979, FE-2893-45, April 1980.

———, Phase V. EDS Product Quality. Interim Report, FE-2893-68, March 1981a.

———, Phase V. Annual Technical Progress Report for the Period July 1, 1980–June 30, 1981, FE-2893-74 (Volume 1), December 1981b.

———, Phase V. Quarterly Technical Progress Report for the Period October–December 1981, FE-2893-83, March 1982.

Finke, H. L., J. F. Messerly, S. H. Lee, A. G. Osborn, and D. R. Douslin, Comprehensive Thermodynamic Studies of Seven Aromatic Hydrocarbons, *J. Chem. Thermodyn.*, **9**, 937 (1977).

Francis, H. E., and W. G. Lloyd, Predicting Heating Values from Elemental Composition, *J. Coal Qual.*, **2**(2), 21 (1983).

Gammon, B. E., J. E. Callanan, I. A. Hossenlopp, A. G. Osborn, and W. D. Good, Heat Capacity, Vapor Pressure, and Derived Thermodynamic Properties of Octahydroanthracene, *Proc. Eighth Symposium Thermophysical Properties*, Vol. II, ASME, New York, 1982, p. 402.

Goursot, P., H. L. Girdhar, and E. F. Westrum, Jr., Thermodynamics of Polynuclear Aromatic Molecules. III. Heat Capacities and Enthalpies of Fusion of Anthracene, *J. Phys. Chem.*, **74**, 2538 (1970).

Gray, J. A., C. J. Brady, J. R. Cunningham, J. R. Freeman, and G. M. Wilson, Thermophysical Properties of Coal Liquids. 1. Selected Physical, Chemical, and Thermodynamic Properties of Narrow Boiling Range Coal Liquids, *Ind. Eng. Chem. Process Des. Dev.*, **22**, 410 (1983).

Gray, J. A., and G. D. Holder, Selected Physical, Chemical, and Thermodynamic Properties of Narrow Boiling Range Coal Liquids from SRC-II Process, Supplementary Property Data, Report No. DOE/ET/10104-44, April 1982.

Grebe, J. J., Diphenyl-Type Compounds for High Temperature Heating, *Chem. Met. Eng.*, **39**(4), 213 (1932).

Helfrey, P. F., D. A. Heiser, and B. H. Sage, Isobaric Heat Capacities at Bubble Point. Two Trimethylbenzenes and *n*-Heptane, *Ind. Eng. Chem.*, **47**, 2385 (1955).

Hougen, O. A., K. M. Watson, and R. A. Ragatz, *Chemical Process Principles. Part I. Material and Energy Balances*, 2nd Ed., Wiley, New York, 1954.

Jain, S. R., and R. Sundararajan, New Method of Calculating Calorific Values from Elemental Compositions of Fossil Fuels, *Fuel*, **60**, 1079 (1981).

Johnson, R. L., and H. G. Grayson, Enthalpy of Petroleum Fractions, *Petroleum Refiner*, **40**(2), 123 (1961).

Joswick, R. L., and P. C. Madden, II, Heat of Reaction in the EDS Coal Liquefaction Process, presented at the 1983 Annual AIChE Meeting, Washington, DC, October 30–November 4, 1983.

Kesler, M. G., and B. I. Lee, Improve Prediction of Enthalpy of Fractions, *Hydrocarbon Process.*, **55**(3), 153 (1976).

Kidnay, A. J., and V. F. Yesavage, Enthalpy Measurement of Coal-Derived Liquids. Quarterly Technical Progress Report for the Period January–March 1977, FE-2035-7, April 1977a.

———, Enthalpy Measurement of Coal-Derived Liquids. Quarterly Technical Progress Report for the Period April–June 1977, FE-2035-8, July 1977b.

———, Enthalpy Measurement of Coal-Derived Liquids. Quarterly Technical Progress Report for the Period July–September 1977, FE-2035-9, October 1977c.

———, Enthalpy Measurement of Coal-Derived Liquids. Quarterly Technical Progress Report for the Period October–December 1977, FE-2035-10, January 1978.

———, Enthalpy Measurement of Coal-Derived Liquids. Final Report for the Period June 1975–March 1979, DOE Contract EX-76-C-01-2035, January 1980a.

———, Enthalpy Measurement of Coal-Derived Liquids. Quarterly Technical Progress Reports for the Period April–June 1979 and July–September 1979, 79ET 13395-1,2, September 1980b.

———, Enthalpy Measurement of Coal-Derived Liquids. Quarterly Technical Progress Report for the Period July–September 1980, 79ET 13395-6, September 1980c.

———, Enthalpy Measurement of Coal-Derived Liquids. Quarterly Technical Progress Report for the Period October–December 1980, 79ET 13395-7, February 1981.

————, Enthalpy Measurement of Coal-Derived Liquids. Technical Progress Report, August 1981–January 1982, DOE/PC/40787-T1, April 1982.

Kudchadker, A. P., and S. A. Kudchadker, *Xylenols*, Key Chemicals Data Books, Texas A & M U., College Station, 1978.

Kudchadker, A. P., S. A. Kudchadker, R. C. Wilhoit, *Phenol*, Key Chemicals Data Books, Texas A & M U., College Station, 1977.

————, *Tetralin*, API Publication 705, 1978a.

————, *Naphthalene*, API Publication 707, 1978b.

————, *Cresols*, Key Chemicals Data Books, Texas A & M U., College Station, 1978c.

Lee, B. I., and M. G. Kesler, A Generalized Thermodynamic Correlation Based on Three Parameter Corresponding States, *AIChE J.*, **21**, 510 (1975).

Lee-Bechtold, S. H., I. A. Hossenlopp, D. W. Scott, A. G. Osborn, and W. D. Good, A Comprehensive Study of 9,10-Dihydrophenanthrene, *J. Chem. Thermodyn.*, **11**, 469 (1979).

Lloyd, W. G., and D. A. Davenport, Applying Thermodynamics to Fossil Fuels, *J. Chem. Ed.*, **57**(1), 56 (1980).

McCullough, J. P., H. L. Finke, J. F. Messerly, S. S. Todd, T. C. Kincheloe, and G. Waddington, The Low-Temperature Thermodynamic Properties of Naphthalene, 1-Methylnaphthalene, 2-Methylnaphthalene, 1,2,3,4-Tetrahydronaphthalene, *trans*-Deca-hydronaphthalene, and *cis*-Decahydronaphthalene, *J. Phys. Chem.*, **61**, 1105 (1957).

Messerly, J. F., S. S. Todd, and H. L. Finke, Low-Temperature Thermodynamic Properties of *n*-Propyl- and *n*-Butylbenzene, *J. Phys. Chem.*, **69**, 4304 (1965).

Mohr, G. D., M. Mohr, A. J. Kidnay, and V. F. Yesavage, The Enthalpies of 2,6-Dimethyl-pyridine and *m*-Cresol between 314 and 669 K and at Pressures to 10.3 MPa, *J. Chem. Thermodyn.*, **15**, 425 (1983).

Mott, R. A., and C. E. Spooner, The Calorific Value of Carbon in Coal: The Dulong Relationship, *Fuel*, **19**, 226, 242 (1940).

Mraw, S. C., J. L. Heidman, S. C. Hwang, and C. Tsonopoulos, Heat Capacity of Coal Liquids, *Ind. Eng. Chem. Process Des. Dev.*, **23**, 577 (1984).

Mraw, S. C., and D. F. O'Rourke, Accuracy of Differential Scanning Calorimetry for Heat Capacities of Organic Compounds at High Temperatures. New Method for Enthalpy of Fusion by D.S.C., *J. Chem. Thermodyn.*, **13**, 199 (1981).

O'Rourke, D. F., and S. C. Mraw, Heat Capacities and Enthalpies of Fusion of Dibenzo-thiophene (220–560 K) and of Biphenyl, Cyclohexylbenzene, and Cyclohexylcyclohexane (220–470 K). Enthalpies and Temperatures of Three Transitions in Solid Cyclohexyl-cyclohexane, *J. Chem. Thermodyn.*, **15**, 489 (1983).

Pitzer, K. S., and D. W. Scott, The Thermodynamics and Molecular Structure of Benzene and Its Methyl Derivatives, *J. Am. Chem. Soc.*, **65**, 803 (1943).

Putnam, W. E., and J. E. Kilpatrick, Entropy, Heat Capacity, and Heats of Transition of 1,2,4-Trimethylbenzene, *J. Chem. Phys.*, **27**, 1075 (1957).

Reid, R. C., J. M. Prausnitz, and T. K. Sherwood, *The Properties of Gases and Liquids*, 3rd Ed., McGraw-Hill, New York, 1977.

Ringen, S., J. Lanum, and F. P. Miknis, Calculating Heating Values from Elemental Compositions of Fossil Fuels, *Fuel*, **58**, 69 (1979).

San José, J. L., G. Mellinger, R. C. Reid, Measurements of the Isobaric Heat Capacity of Liquids and Certain Mixtures Above the Normal Boiling Point, *J. Chem. Eng. Data*, **21**, 114 (1976).

Schlinger, W. G., and B. H. Sage, Isobaric Heat Capacities at Bubble Point. *cis*-Butene-2, Isopropylbenzene, and *n*-Decane, *Ind. Eng. Chem.*, **44**, 2454 (1952).

Scott, D. W., G. B. Guthrie, J. F. Messerly, S. S. Todd, W. T. Berg, I. A. Hossenlopp, and J. P. McCullough, Toluene: Thermodynamic Properties, Molecular Vibrations, and Internal Rotation, *J. Phys. Chem.*, **66**, 911 (1962).

Scott, R. B., and F. G. Brickwedde, Thermodynamic Properties of Solid and Liquid Ethylbenzene from 0 to 300 K, *J. Res. Nat. Bur. Std.*, **35**, 501 (1945).

Sharma, R., Enthalpy Measurements for a Coal-Derived Naphtha and Middle Distillate and Characterization of Coal Liquids for Their Extent of Association, Ph.D. dissertation, Colorado School of Mines, 1980.

Smith, N. K., S. H. Lee-Bechtold, and W. D. Good, Thermodynamic Properties of Materials Derived from Coal Liquefaction, DOE/BETC Technical Progress Report TPR-79/2, January 1980.

Starling, K. E., M. R. Brulé, C. T. Lin, and S. Watanasiri, Calculation of Distillable-Coal-Fluid Thermophysical Properties Using Multiparameter Corresponding-States Correlations, Coal-Calc Project Report OU/IGT/S-14366/1, University of Oklahoma, Norman, 1980.

Stull, D. R., E. F. Westrum, Jr., and G. C. Sinke, *The Chemical Thermodynamics of Organic Compounds*, Wiley, New York, 1969.

Svoboda, V., V. Charvatova, V. Majer, and V. Hynek, Determination of Heats of Vaporization and Some Other Thermodynamic Properties for Four Substituted Hydrocarbons, *Coll. Czech. Chem. Commun.*, **47**, 543 (1981).

Swanson, A. C., and C. F. Chueh, Estimation of Liquid Heat Capacity by the Halcon Physical Data System, presented at the 65th Annual AIChE Meeting, New York, November 1972.

Taylor, R. D., B. H. Johnson, and J. E. Kilpatrick, Entropy, Heat Capacity, and Heats of Transition of 1,2,3-Trimethylbenzene, *J. Chem. Phys.*, **23**, 1225 (1955).

Watson, K. M., and E. F. Nelson, Improved Methods for Approximating Critical and Thermal Properties of Petroleum Fractions, *Ind. Eng. Chem.*, **25**, 880 (1933).

Wieczorek, S. A., and R. Kobayashi, Heats of Vaporization of Five Polynuclear Aromatic Compounds at Elevated Temperatures, *J. Chem. Eng. Data*, **26**, 11 (1981).

Chapter 7

LIQUID DENSITY

Accurate predictions of the density of liquid mixtures are important in the design and specification of most processing equipment in coal conversion plants. Furthermore, accurate predictions of liquid densities must extend to high temperatures (up to 900°F) and high pressures (up to 3000 psia) for most of the conversion processes currently under development.

Although very little information exists on the density of either petroleum-derived or coal-derived liquids at these conditions, recent investigations by Hwang et al. (1982) and Gray et al. (1983) have resulted in a substantial body of valuable data on the density of coal-liquid systems. These data, some of which extend to temperatures above 800°F and pressures greater than 3000 psia, provide the basis for the evaluation of several correlations for predicting the density of coal-liquid mixtures.

In this chapter, we review several well-established correlations for the saturated liquid density of hydrocarbons and evaluate their ability to predict the liquid density of coal-derived fluids. The effect of pressure on the liquid density is also investigated. Finally, we establish mixing rules for the various correlations in order to calculate the density of coal-liquid mixtures, including those containing supercritical components. The developed procedures are then evaluated with the available coal-liquid density data.

SATURATED LIQUID DENSITY CORRELATIONS

In their evaluation of the density data for EDS liquids with and without dissolved hydrogen, Hwang et al. (1982) used a form of the Riedel equation (Riedel, 1954).

$$\rho_r = \frac{\rho}{\rho_c} = 1 + 0.85(1 - T_r) + (1.6916 + 0.9846\omega)(1 - T_r)^{1/3} \qquad (7.1)$$

where ρ_r is the reduced density (density/critical density), T_r is the reduced temperature (T/T_c), and ω is the acentric factor. For these comparisons, ρ_c is determined from the specific gravity at 60/60°F via the Watson relationship (Watson, 1943)

$$\rho = \rho_r \left(\frac{\rho_{\text{ref}}}{\rho_{r,\text{ref}}} \right) \tag{7.2}$$

where ρ_{ref} is the density at 60°F, $\rho_{r,\text{ref}}$ is obtained via Eq. 7.1 such that T_r is equal to $(459.67 + 60)/T_c$, and ρ_r is equal to 1. Note that the mass density at 60°F is given by the specific gravity according to $\rho_{60} = 0.999024S$. The parametric behavior of the Riedel correlation is illustrated in Figure 7.1.

Gray et al. (1983) used a form of the Rackett equation (Rackett, 1970) to smooth their liquid density data. The general form of the Rackett equation is

$$\rho = \left(\frac{P_c}{RT_c} \right) (Z_{\text{RA}})^{-[1+(1-T_r)^{2/7}]} \tag{7.3}$$

Z_{RA}, which corresponds closely to the critical compressibility factor, is treated as a specified constant for each fraction and is obtained from the

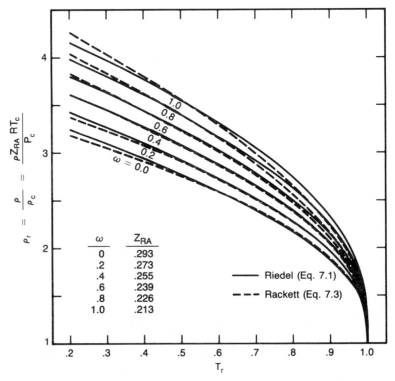

Figure 7.1   Riedel and Rackett correlations for saturated liquid density.

density at 60°F by

$$\log Z_{RA} = \frac{\log(\rho R T_c / P_c)}{-[1 + (1 - T_{r,60°F})^{2/7}]} \tag{7.4}$$

where ρ is the molar density at 60°F and $T_{r,60°F}$ is the reduced temperature corresponding to 60°F for the coal-liquid fraction. The parametric behavior of the Rackett equation is shown in Figure 7.1 along with that of the Riedel equation.

In a subsequent paper, Holder and Gray (1983) developed a one-constant modification of the Rackett equation, based on the data reported by Gray et al. (1983) for SRC-II liquids. This equation is given by

$$\rho_r = \frac{\rho}{\rho_c} = (0.28412)^{-(1-T_r)^{2/7}} \tag{7.5}$$

where ρ_c is obtained from the specific gravity at 60/60°F, and the critical temperature is given by Eq. 3.12 (Brulé et al., 1982). Although Eq. 7.5 reproduced the SRC-II liquid density data with errors less than $\pm 1\%$, it cannot be generalized to mixtures of coal-liquid fractions and defined compounds, such as the coal-liquids/hydrogen mixtures for which Hwang et al. (1982) have reported liquid density data.

In addition to the Riedel correlation, Eq. 7.1, and the Rackett correlation, Eq. 7.3, the recently developed COSTALD correlation (Hankinson and Thomson, 1979) is examined for its applicability to coal-liquid mixtures. The COSTALD correlation for saturated liquid densities is given by

$$\rho = \left(\frac{1}{v^*}\right)\left(\frac{1}{v_r^{(o)}}\right)\frac{1}{(1 - \omega_{SRK} v_r^{(\delta)})} \tag{7.6a}$$

where

$$v_r^{(o)} = 1 - 1.52816(1 - T_r)^{1/3} + 1.43907(1 - T_r)^{2/3}$$
$$- 0.81446(1 - T_r) + 0.190454(1 - T_r)^{4/3} \tag{7.6b}$$

$$v_r^{(\delta)} = (-0.296123 + 0.386914 T_r - 0.0427258 T_r^2$$
$$- 0.0480645 T_r^3)/(T_r - 1.00001) \tag{7.6c}$$

and ω_{SRK} is an acentric factor value derived from the Soave-Redlich-Kwong equation of state. However, for these comparisons, the acentric factor obtained from Eq. 4.15 was used in place of ω_{SRK}. The characteristic volume v^*, which has been shown by Hankinson and Thomson (1979) to correspond closely to the critical volume, is treated as a specified constant for a given compound. In this evaluation, v^* is obtained from the coal-liquid fraction specific gravity at 60/60°F in a manner analogous to that previously described for the Riedel correlation ρ_c and the Rackett equation Z_{RA} parameters. Furthermore, this approach should be used to establish the critical density (volume, compressibility factor) for the coal-liquid fractions, since it results in a critical density value which is consistent with the liquid density predictions.

EFFECT OF PRESSURE—ISOTHERMAL COMPRESSIBILITY

At pressures greater than saturation, the densities of liquids increase with increasing pressure at constant temperature. The isothermal compressibility, β, is defined as

$$\beta = -\frac{1}{v}\left(\frac{\partial v}{\partial P}\right)_T = \frac{1}{\rho}\left(\frac{\partial \rho}{\partial P}\right)_T \tag{7.7}$$

where ρ is the molar density. Equation 7.7 is typically the basis for correlations of the effect of pressure on liquid density. Alternatively, the Watson relationship, Eq. 7.2, can be restated as

$$\rho = \left(\frac{K}{K_{ref}}\right)\rho_{ref} \tag{7.8}$$

where ρ and ρ_{ref} are the desired density and the density at some reference condition, respectively, and K and K_{ref} are values of the corresponding correlation parameters and are generally functions of T_r and P_r. Both approaches have been used successfully to correlate the effect of pressure on liquid density.

In their analysis, Hwang et al. (1982) made use of Lydersen's tabular correlation (Lydersen et al., 1955) for predicting the liquid density at pressures greater than saturation. The Lydersen tabulation provides the reduced density, ρ_r, as a function of T_r, P_r, and critical compressibility factor, Z_c, which may be obtained from the following approximate relationship

$$Z_c = \frac{1}{3.41 + 1.28\omega} \tag{7.9}$$

Thus from the Riedel correlation, Eq. 7.1, the Lydersen correlation, and the Watson relationship, Eq. 7.2, the liquid density at reduced temperatures from 0.30 to 1.00 and reduced pressures up to 30 could be predicted. Their analysis demonstrated that this approach accurately predicted the liquid densities of EDS coal liquids (with and without dissolved hydrogen).

Two alternatives to the Lydersen correlation for predicting the effect of pressure on liquid density are the correlations of Chueh and Prausnitz (1969), which incorporates the isothermal compressibility β, and the correlation presented by Rea et al. (1973), which is an analytical representation of the Lu (1959) charts as revised by Ewbank and Harden (1967). The Chueh-Prausnitz correlation is given by

$$\left(\frac{\rho}{\rho^s}\right) = [1 + 9\beta(P - P^s)]^{1/9} \tag{7.10a}$$

$$\beta = \frac{v_c}{RT_c}(1 - 0.89\sqrt{\omega})\exp(6.9547 - 76.2853\,T_r$$

$$+ 191.306\,T_r^2 - 203.5472\,T_r^3 + 82.7631\,T_r^4) \tag{7.10b}$$

where ρ^s and P^s are the saturation density and pressure, respectively, and v_c is the critical volume. Equation 7.10 is valid for reduced temperatures ranging from 0.40 to 0.98. The effect of pressure on liquid density via Eq. 7.10 is illustrated in Figure 7.2 for a fluid with $\omega = 0.35$ (e.g., tetralin).

The correlation reported by Rea et al. (1973) is of the form

$$K = A_0 + A_1 T_r + A_2 T_r^2 + A_3 T_r^3 \tag{7.11a}$$

$$A_i = B_{0i} + B_{1i}P_r + B_{2i}P_r^2 + B_{3i}P_r^3 + B_{4i}P_r^4 \tag{7.11b}$$

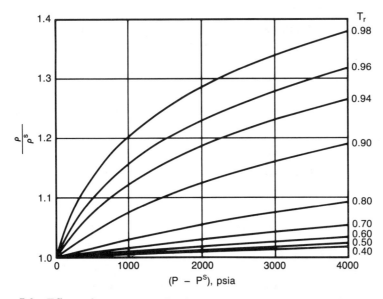

Figure 7.2 Effect of pressure on liquid density with Chueh-Prausnitz correlation for $\omega = 0.35$.

where K is the correlation parameter for the Watson relationship, Eq. 7.8. The values of the coefficients B_{ji} are given below.

i	B_{0i}	B_{1i}	$B_{2i}(10^3)$	$B_{3i}(10^5)$	$B_{4i}(10^6)$
0	1.6368	−0.04615	2.1138	−0.7845	−0.6923
1	−1.9693	0.21874	−8.0028	8.2823	5.2604
2	2.4638	−0.36461	12.8763	14.8059	−8.6895
3	−1.5841	0.25136	−11.3805	9.5672	2.1812

Finally, the COSTALD correlation includes the effect of pressure given by a modified form of the Tait equation

$$\rho = \rho^s \left[1 - C \ln\left(\frac{B+P}{B+P^s}\right) \right]^{-1} \tag{7.12a}$$

where

$$\frac{B}{P_c} + 1 = -9.070217(1 - T_r)^{1/3} + 62.45326(1 - T_r)^{2/3}$$
$$- 135.1102(1 - T_r) + e(1 - T_r)^{4/3} \tag{7.12b}$$

$$e = \exp(4.79594 + 0.250047\omega + 1.14188\omega^2) \tag{7.12c}$$

$$C = 0.0861488 + 0.0344483\omega \tag{7.12d}$$

All of the correlations described above for predicting the effect of pressure on liquid density have generally been developed from data on light hydrocarbons and compounds typically found in petroleum-derived fluids. Therefore, their applicability to coal-liquid-containing mixtures needs to be demonstrated.

In order to do so, the saturation pressure and mixture characterization constants need to be determined. For a liquid mixture, the saturation pressure, P^s, is estimated from the generalized Riedel correlation, Eq. 4.11, from the mixture pseudocritical temperature and pressure, T_{cm} and P_{cm}, and the mixture acentric factor, ω_m. The calculation of these and other required mixture parameters is described in the next section.

MIXING RULES FOR LIQUID DENSITY PREDICTIONS

In order to calculate the liquid density of coal liquid-containing mixtures, mixing rules are needed for the calculation of the appropriate pseudocritical constants. Recent investigations (Hankinson and Thomson, 1979; Spencer and Danner, 1973; Veeranna and Rihani, 1980) have examined several potential combination rules for calculating T_{cm}, P_{cm}, Z_{cm}, and ω_m for use in predicting liquid densities. The results of these investigations are summarized in this section for each of the saturated liquid density correlations under consideration.

For both the Riedel and Rackett correlations, Eqs. 7.1 and 7.3, the mixture pseudocritical temperature is predicted by the correlation proposed by Chueh and Prausnitz (1967)

$$T_{cm} = \sum_i \sum_j \Phi_i \Phi_j T_{cij} \tag{7.13a}$$

$$T_{cij} = 8 \left[\frac{(v_{ci}^{1/3} v_{cj}^{1/3})^{1/2}}{(v_{ci}^{1/3} + v_{cj}^{1/3})} \right]^3 (T_{ci} T_{cj})^{1/2} \tag{7.13b}$$

$$\Phi_i = \frac{x_i v_{ci}}{\sum_j x_j v_{cj}} \tag{7.13c}$$

where Φ_i is the critical volume fraction of component i and v_{ci} is its critical volume. For both the Riedel and Rackett correlations, the value of v_{ci} for the coal-liquid fractions is determined from the correlation critical parameter (ρ_c for Riedel, Z_{RA} for Rackett) as obtained from the specific gravity at 60/60°F. As mentioned previously, this procedure should be used to determine the critical density (volume) of the coal-liquid fractions.

The COSTALD correlation uses mixing rules for T_{cm} similar to those utilized by Lee and Kesler (1975) in their corresponding-states correlation for the PVT and thermodynamic properties of nonpolar mixtures. These are given by

$$T_{cm} v_m^* = \sum_i \sum_j x_i x_j T_{cij} v_{ij}^* \tag{7.14a}$$

$$T_{cij} v_{ij}^* = (T_{ci} T_{cj})^{1/2} (v_i^* v_j^*)^{1/2} \tag{7.14b}$$

$$v_m^* = \frac{1}{8} \sum_i \sum_j x_i x_j (v_i^{*1/3} + v_j^{*1/3})^3 \tag{7.14c}$$

where v_i^* is the characteristic volume of component i. As discussed previously, v_i^* corresponds closely to the critical volume and is obtained from the coal-liquid fraction specific gravity at 60/60°F.

The pseudocritical pressure, P_{cm}, is needed to obtain an estimate of the saturation pressure from Eq. 4.11 and to calculate the mixture reduced pressure, P/P_{cm}. Both of these values are needed to predict the effect of pressure on the mixture liquid density. For both the Riedel and Rackett correlations, P_{cm} is obtained from

$$\frac{1}{P_{cm}} = \frac{1}{T_{cm}} \sum_i x_i \frac{T_{ci}}{P_{ci}} \tag{7.15}$$

where T_{cm} is the mixture pseudocritical temperature obtained from Eq. 7.13. The COSTALD correlation uses a pseudocritical pressure obtained from

$$P_{cm} = Z_{cm}RT_{cm}/v_m^* \qquad (7.16a)$$

where T_{cm} and v_m^* are given by Eq. 7.14 and Z_{cm} is given by

$$Z_{cm} = 0.291 - 0.080\omega_m \qquad (7.16b)$$

and ω_m, the mixture acentric factor, is given by

$$\omega_m = \sum_i x_i\omega_i \qquad (7.17)$$

Equation 7.17 is used for all of the liquid density correlations under consideration.

The mixing rules for the Riedel correlation critical density and the Rackett correlation "critical" compressibility parameter, Z_{RAm}, are given by the following relationships. For ρ_{cm} in Eq. 7.1,

$$\frac{1}{\rho_{0m}} = \sum_i \frac{x_i}{\rho_{ci}(1.85 + 1.6916 + 0.9846\omega_i)} \qquad (7.18a)$$

$$\rho_{cm} = \frac{\rho_{0m}}{1.85 + 1.6916 + 0.9846\omega_m} \qquad (7.18b)$$

where ρ_{0m} is the hypothetical liquid mixture density at 0°R as given by the Riedel equation and assuming ideal volumetric mixing. As previously discussed, ρ_{ci} is the critical density of component i obtained from the specific gravity at 60/60°F. Finally, for the Rackett correlation, Eq. 7.3, the value of Z_{RAm} is given by

$$Z_{RAm} = \sum_i x_i Z_{RAi} \qquad (7.19)$$

With these mixing rules established, the liquid density correlations described above can be evaluated for their ability to predict the density of coal-liquid-containing mixtures.

DATA EVALUATION PROCEDURES

The liquid density correlations investigated in this chapter require that the specific gravity and boiling point of the coal-liquid fractions be known. These two parameters are used to predict other "characterization" properties that are calculated by the correlations discussed and recommended in Chapters 3 and 4. These properties include critical temperature (Eq. 3.10), critical pressure (Eq. 3.15), and acentric factor (from the generalized Riedel correlation and the boiling point according to Eq. 4.15). In addition, if the molecular weight is needed, it is calculated by Eq. 3.5.

The density data reported by Hwang et al. (1982) and Gray et al. (1983) include sufficient information regarding the characterization of the coal-liquid fractions. The data reported by Gray et al. are mostly for relatively narrow-boiling (< 50°F range) fractions for which the mid-range boiling point and specific gravity at 60/60°F are reported. The data reported by Hwang et al., however, are for very wide-boiling fractions (up to 700°F range) which must be represented by several narrow-boiling range fractions in order to obtain reasonable liquid density predictions.

Therefore, each of the coal liquids reported by Hwang et al. was represented by 10 fractions, obtained by plotting the gas chromatographic distillation data (temperature versus cumulative weight percent off; see Appendix A) and then dividing the smoothed distillation curve into 10 wt% increments. The mid-range temperature of each "fraction" was taken as the normal boiling point. A specific gravity for each fraction was calculated from the boiling point via Eq. 3.4 and then adjusted to match the reported total liquid specific gravity according to

$$\frac{1}{S_m} = \sum_i \frac{w_i}{S_i} \qquad (7.20)$$

where w_i is the weight fraction of component i. The adjustments of the individual fraction specific gravities were usually less than 2%. The resulting characterization of the narrow fractions is given in Appendix B.

Once the boiling point, specific gravity, and other correlation parameters are established for each coal-liquid fraction, the liquid density correlations can be compared with the available experimental data. Since these data were generally measured at pressures greater than saturation, the evaluation involves calculating the liquid density at the bubble-point (i.e., saturation) pressure via correlations for saturated liquid density, and then accounting for the effect of pressure on the liquid density.

ANALYSIS OF COAL LIQUID DATA

As mentioned at the beginning of the chapter, the basis for the evaluation of the previously described liquid density correlations is the recent and extensive investigations of the thermophysical properties of coal liquids by Gray et al. (1983) and by Hwang et al. (1982). These two sources report about 450 liquid density measurements for temperatures up to 800°F and pressures ranging from near saturation to over 3000 psia. These data were obtained for both narrow-boiling (generally less than 50°F-wide SRC-II cuts boiling up to about 750°F) and wide-boiling coal-liquid fractions (about 300 to 700°F boiling range EDS cuts). Finally, about 40% of the density data for the wide-boiling range liquids include dissolved hydrogen, thus providing an additional test of both the density correlations and the mixing rules.

The saturated liquid density and compressibility correlations were evaluated in the following manner. The Riedel correlation, Eq. 7.1, was analyzed in combination with both the Chueh-Prausnitz (Eq. 7.10) and the Rea et al. (Eq. 7.11) correlations for the effect of pressure on liquid density. Similarly, the Rackett correlation, Eq. 7.3, was evaluated with both Eqs. 7.10 and 7.11 for the effect of pressure on density. Finally, the COSTALD correlation, Eq. 7.6, was considered only with the modified Tait equation, Eq. 7.12. Therefore, five correlation combinations were evaluated for their ability to predict the density of coal-liquid mixtures.

The analysis of these data is summarized in Tables 7.1 and 7.2 for the SRC-II fractions reported by Gray et al. (1983) and the EDS fractions reported by Hwang et al. (1982), respectively. Table 7.1 shows that all of

Table 7.1   Analysis of SRC-II Coal Liquid Densities

SRC Liquid[b]	No. Data Points	Average Absolute Deviations/Bias[a]				
		RCP[c]	RR	RACP	RAR	COSTALD
1	6	3.34/−3.34	3.29/−3.29	3.03/−3.03	2.98/−2.98	1.59/−1.41
2	6	1.65/−0.30	1.63/−0.20	1.71/−0.33	1.69/−0.23	1.94/ 1.44
3	6	1.40/−1.40	1.22/−1.22	1.51/−1.51	1.33/−1.32	0.62/ 0.29
4	6	1.17/−1.10	0.99/−0.92	1.35/−1.27	1.17/−1.10	0.37/ 0.37
5	6	0.52/−0.52	0.47/−0.47	0.64/−0.64	0.60/−0.60	0.62/ 0.62
6	6	0.32/−0.32	0.20/−0.11	0.36/−0.36	0.19/−0.15	0.62/ 0.62
7	6	0.89/−0.88	0.84/−0.84	0.97/−0.96	0.93/−0.93	0.24/−0.07
8	6	0.32/−0.32	0.18/−0.14	0.44/−0.44	0.26/−0.26	0.40/ 0.36
9	6	0.21/−0.08	0.19/−0.07	0.33/−0.24	0.28/−0.22	0.92/ 0.92
10	6	0.71/−0.71	0.70/−0.70	0.87/−0.87	0.86/−0.86	0.20/ 0.19
11	6	0.21/−0.21	0.20/−0.20	0.34/−0.34	0.33/−0.33	0.58/ 0.58
12	6	0.30/ 0.24	0.31/ 0.24	0.25/ 0.18	0.26/ 0.18	0.84/ 0.84
13	6	0.52/ 0.52	0.51/ 0.51	0.47/ 0.47	0.46/ 0.46	1.04/ 1.04
15	6	0.25/−0.21	0.26/−0.23	0.27/−0.25	0.29/−0.27	0.50/ 0.34
16	6	0.41/−0.37	0.43/−0.39	0.43/−0.41	0.45/−0.43	0.54/ 0.06
Total	90	0.81/−0.60	0.76/−0.54	0.86/−0.67	0.81/−0.60	0.73/ 0.41

[a] AAD = [$\sum$ |calc − exp|/exp] × 100/N; Bias = [$\sum$ (calc − exp)/exp] × 100/N; N is number of data points.
[b] Listed in increasing boiling-point order (Gray et al., 1983).
[c] RCP: Riedel with Chueh-Prausnitz; RR: Riedel with Rea et al.; RACP: Rackett with Chueh-Prausnitz; RAR: Rackett with Rea et al.; COSTALD: COSTALD correlation framework.

Table 7.2   Analysis of EDS Coal Liquid Densities

EDS Liquid	No. Data Points	Average Absolute Deviations/Bias[a]				
		RCP[a]	RR	RACP	RAR	COSTALD
Hydrogen-Free Coal Liquids						
IA-3	30	1.32/ 1.32	2.35/ 2.22	1.22/ 1.22	2.22/ 2.12	2.82/ 2.82
IA-6	36	0.55/ 0.55	1.82/ 1.50	0.22/ 0.01	1.31/ 0.96	1.93/ 1.93
IA-10	32	0.41/ 0.39	1.68/ 1.52	0.44/−0.42	1.06/ 0.69	1.94/ 1.94
IHS	33	0.73/ 0.03	0.85/ 0.60	0.88/−0.60	0.71/−0.05	1.33/ 1.27
WA-5	30	0.85/−0.85	1.22/ 0.07	1.26/−1.26	1.08/−0.34	0.77/ 0.56
WA-6	26	1.28/−1.28	1.04/−0.64	1.53/−1.53	1.07/−0.89	0.35/ 0.05
WV-1	29	0.84/−0.84	1.25/−0.16	1.12/−1.12	1.15/−0.43	0.81/ 0.50
Total	216	0.83/−0.05	1.47/ 0.79	0.92/−0.49	1.23/ 0.34	1.46/ 1.35
Hydrogen-Containing Coal Liquids						
IA-3	21	3.04/ 3.04	4.20/ 3.87	1.40/ 0.19	1.98/ 0.98	3.83/ 3.83
IA-6	21	2.76/ 2.76	4.14/ 3.94	1.76/−1.13	1.72/−0.02	3.68/ 0.46
WA-2	22	0.96/−0.96	1.25/−0.94	2.63/−2.63	2.62/−2.62	0.53/−0.33
WA-5	22	2.46/−2.46	1.66/−1.66	5.45/−5.45	4.71/−4.71	2.63/−2.63
WA-6	30	1.17/−1.17	1.29/−0.35	2.35/−2.35	1.55/−1.55	0.43/−0.43
WV-1	28	1.40/−1.40	1.29/−0.44	2.60/−2.60	1.66/−1.66	1.75/ 0.00
Total	144	1.88/−0.19	2.18/ 0.58	2.69/−2.37	2.31/−1.63	1.75/ 0.00
Summary (SRC-II, from Table 7.1, and EDS)						
Total	450	1.16/−0.20	1.56/ 0.46	1.47/−1.13	1.49/−0.48	1.41/ 0.73
Without H₂	306	0.82/−0.21	1.26/ 0.40	0.90/−0.54	1.11/ 0.06	1.25/ 1.07

[a]See Table 7.1.

the correlations perform equally well for the SRC-II liquids, with the COSTALD correlation exhibiting the lowest overall average deviation and bias in the density predictions. However, since the stated accuracy of these measurements is generally 1 to 2%, all of the correlations perform within the experimental uncertainty and are, therefore, essentially equivalent.

Table 7.1 also shows that the largest deviations exist for the lowest boiling liquid fractions and that the errors decrease for the higher boiling fractions. This occurs because the highest temperature investigated by Gray et al. for these liquids was about 440°F, which is very close to the critical temperature for the lightest fractions, but well removed from the critical region for the heavier cuts. In fact, the larger errors for the lightest fractions are due to errors in the predicted liquid density close to the critical temperature. Furthermore, these errors do not necessarily reflect inadequacies in the liquid density correlations, but probably indicate that the predicted value of the critical temperature is inconsistent with the experimental liquid density data.

Figure 7.3 illustrates this behavior for the SRC-II fraction No. 1, where the data suggest a critical temperature near 500°F, whereas the Riedel correlation, Eq. 7.1, predicts liquid density behavior consistent with the critical temperature predicted with Eq. 3.10, 472°F. In addition, the specific gravity at 60/60°F, 0.7234, appears to be low relative to the data at higher temperatures, thus further biasing the density predictions. The magnitude of the deviations shown here is probably typical of errors to be expected as the

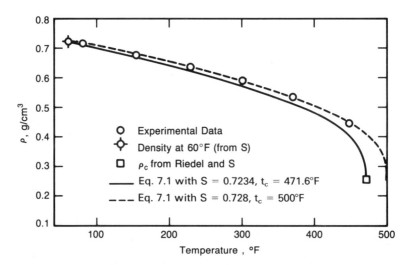

Figure 7.3   Saturated liquid density predictions for SRC-II Coal Liquid No. 1 with the Riedel equation.

critical region is approached, although the deviations may be either positive or negative.

Table 7.2 summarizes the analysis of the density data for the wide-boiling EDS fractions reported by Hwang et al. (1982) for mixtures without and with dissolved hydrogen. For the hydrogen-free systems, the Riedel/Chueh-Prausnitz correlations appear to provide the most accurate liquid density predictions (0.83% AAD) with essentially no bias (−0.05%). The Rackett/Chueh-Prausnitz correlations exhibit similar errors, 0.93% AAD, but with a small negative (−0.49%) bias. These errors compare favorably with the results obtained by Hwang et al. (1982) with the Riedel/Lydersen correlations. Their analysis resulted in an average absolute error of about 1.4% and a bias of +0.42% for their EDS data. Finally, because of the wide-boiling range of EDS liquids under investigation, the density data did not appear to approach the mixture critical region and, thus, the related problems were not observed.

However, Table 7.2 does show that accuracy of liquid density predictions decreases significantly with the introduction of dissolved hydrogen. The COSTALD correlation provides the most accurate predictions for the hydrogen-containing systems, although the Riedel/Chueh-Prausnitz correlations are only marginally less accurate. These larger errors for the hydrogen-containing systems most likely reflect the limitations of the mixing rules associated with each of the liquid density correlations. Nevertheless, average deviations for hydrogen-containing systems are only about 2%, which is nearly equal to the experimental uncertainty.

Finally, the overall errors for coal-liquid mixtures both with and without hydrogen are also summarized in Table 7.2 and show that all of the correlations investigated predict liquid densities with average deviations of less than 1.5%. The most accurate of the combinations appears to be the Riedel correlation, Eq. 7.1, for saturated liquid density with the Chueh-Prausnitz correlation, Eq. 7.10, for the isothermal compressibility. This correlation "framework" predicts liquid densities for both types of systems with deviations of about ±1% over the temperature and pressure ranges considered.

CONCLUSIONS AND RECOMMENDATIONS

All of the correlations evaluated for predicting the liquid density of coal-liquid mixtures appear to be satisfactory. Overall average deviations of 2% or less can be expected for both hydrogen-free and hydrogen-containing mixtures over wide ranges of temperature and pressure. The accuracy of such predictions will most likely decrease rapidly as the critical temperature is approached, because of the uncertainties in the predicted component critical temperatures and the mixing rules for determining T_{cm}.

Furthermore, this analysis demonstrates that the liquid density of coal liquids does not appear to be strongly dependent on the type of coal or the liquefaction process. Liquids from various coal sources (Illinois, Wyoming for EDS, Pittsburgh A-Seam for SRC-II) and liquefaction processes exhibit similar deviations in liquid density predictions.

The best overall predictions were obtained from the Riedel correlation, Eq. 7.1, for the saturated liquid density, in combination with the Chueh-Prausnitz correlation, Eq. 7.10, for the effect of pressure on liquid density. Using these correlations and the associated mixing rules for pseudocritical constants, overall average errors of about ±1% can be expected for coal-liquid mixtures away from the critical region. As the critical temperature is approached (within 50 to 75°F), errors in liquid density can easily exceed 10%.

REFERENCES

Brulé, M. R., C. T. Lin, L. L. Lee, and K. E. Starling, Multiparameter Corresponding States Correlation of Coal-Fluid Thermodynamic Properties, *AIChE J.*, **28**, 616 (1982).

Chueh, P. L., and J. M. Prausnitz, Vapor-Liquid Equilibria at High Pressures: Calculation of Critical Temperatures, Volumes, and Pressures of Nonpolar Mixtures, *AIChE J.*, **13**, 1107 (1967).

———, A Generalized Correlation for the Compressibilities of Normal Liquids, *AIChE J.*, **15**, 471 (1969).

Ewbank, W. J., and D. G. Harden, Correlation Method for Determining Compressed Liquid Densities, *J. Chem. Eng. Data*, **12**, 363 (1967).

Gray, J. A., C. J. Brady, J. R. Cunningham, J. R. Freeman, and G. M. Wilson, Thermophysical Properties of Coal Liquids. 1. Selected Physical, Chemical, and Thermodynamic Properties of Narrow Boiling Range Coal Liquids, *Ind. Eng. Chem. Process Des. Dev.*, **22**, 410 (1983).

Hankinson, R. W., and G. H. Thomson, Calculate Liquid Densities Accurately, *Hydrocarbon Process.*, **58**(9), 277 (1979).

Holder, G. D., and J. A. Gray, Thermophysical Properties of Coal Liquids. 2. Correlating Coal Liquid Densities, *Ind. Eng. Chem. Process Des. Dev.*, **22**, 424 (1983).

Hwang, S. C., C. Tsonopoulos, J. R. Cunningham, and G. M. Wilson, Density, Viscosity and Surface Tension of Coal Liquids at High Temperatures and Pressures, *Ind. Eng. Chem. Process Des. Dev.*, **21**, 127 (1982).

Lee, B. I., and M. G. Kesler, A Generalized Thermodynamic Correlation Based on Three Parameter Corresponding States, *AIChE J.*, **21**, 510 (1975).

Lu, B. C. Y., Estimate Specific Liquid Volumes, *Chem. Eng.*, **66**(9), 137 (1959).

Lydersen, A. L., R. A. Greenkorn, and O. A. Hougen, Generalized Thermodynamic Properties of Pure Fluids, *Univ. Wisconsin Eng. Exp. Sta. Report*, **4**, Madison, October 1955.

Rackett, H. G., Equation of State for Saturated Liquids, *J. Chem. Eng. Data*, **15**, 514 (1970).

Rea, H. E., C. F. Spencer, and R. P. Danner, Effect of Pressure on the Liquid Densities of Pure Hydrocarbons, *J. Chem. Eng. Data*, **18**, 227 (1973).

Riedel, L., Liquid Density in the Saturated State, *Chem.-Ing.-Tech.*, **26**, 259 (1954), in German.

Spencer, C. F., and R. P. Danner, Prediction of Bubble-Point Density of Mixtures, *J. Chem. Eng. Data*, **18**, 230 (1973).

Veeranna, D., and D. N. Rihani, Review of Density Estimation of Saturated Liquid Mixtures, *J. Chem. Eng. Data*, **25**, 267 (1980).

Watson, K. M., Thermodynamics of the Liquid State, *Ind. Eng. Chem.*, **35**, 398 (1943).

Chapter 8

SURFACE TENSION

The surface tension of vapor-liquid mixtures at equilibrium is a property of the interface of the two phases and, much like density, is important in correlations for the design of process equipment such as distillation columns and separator drums. Although these correlations are generally less sensitive to errors in the predicted surface tension values than, say, errors in liquid density, reasonably accurate predictions of surface tension are required to the same high temperatures ($\sim$900°F) and high pressures ($\sim$3000 psia) for which accurate density predictions are needed.

Unlike liquid density, there are essentially no surface tension data for either model compounds or petroleum- or coal-derived fluids at high temperatures. Furthermore, although there are substantial data for model compounds at low to moderate temperatures, there are very few compounds for which data are available at temperatures above the compound's normal boiling point. Recent investigations by Hwang et al. (1982), Gray and Holder (1982), and Gray et al. (1983) have provided data on the surface tension of coal-liquid-containing mixtures up to 700°F and 2000 psia. However, more high-temperature surface tension data are needed on both model compounds and well-defined mixtures of model compounds.

The correlations for model compounds and narrow-boiling fractions are examined first, followed by the recommended method for coal-liquid mixtures. Then this method is applied to the analysis of data for EDS and SRC-II coal-liquid mixtures with H_2 or methane. Generally, agreement with these data is found to be unsatisfactory.

CORRELATIONS FOR SURFACE TENSION OF PURE COMPOUNDS

An extensive review by Reid et al. (1977) shows that two functional forms are generally used to correlate the surface tension of pure fluids. The first is given by

$$\sigma = \sigma_0(1 - T_r)^b \tag{8.1}$$

where σ is the surface tension (typically given in dynes per centimeter) and σ_0 and b are adjustable parameters. For most nonpolar fluids, the value of b

is about 11/9, or 1.222. In addition, several correlations have been proposed for predicting σ_0, the hypothetical surface tension at 0°R. These are described by Reid et al. (1977).

Equation 8.1 has also been used as the basis for corresponding states correlations proposed by Rice and Teja (1982) and Murad (1983). Both correlations provide accurate predictions of the surface tension for pure, nonpolar compounds and reasonable predictions for binary mixtures of nonpolar compounds that are below their critical temperature. However, as is discussed in a later section, these correlations may not be applicable to mixtures containing supercritical components, a serious limitation for process design calculations.

The second formulation for correlating surface tension was first suggested by Macleod (1923) and later developed by Sugden (1924) and Quayle (1953). It is given by

$$\sigma^{0.25} = \frac{P^*}{M}(\rho^L - \rho^V) \tag{8.2}$$

where ρ^L and ρ^V are the saturated liquid and vapor mass densities, respectively, M is the molecular weight, and the constant P^* is the component parachor. Quayle (1953) showed that the parachor could be predicted by a group-contribution procedure for a wide variety of defined compounds, including alcohols, ketones, and ethers. Given an accurate value of the parachor, P^* (or P^*/M), and accurate densities (particularly liquid densities, since ρ^V is negligible below the normal boiling point), the surface tension of a pure component can be predicted to within 2–3% from the triple point to near the critical point, where the surface tension goes to zero.

Figure 8.1 illustrates the temperature dependence of the surface tension for three model compounds—tetralin, 1-methylnaphthalene, and quinoline—at temperatures well below their normal boiling points (no data could be found at higher temperatures). These data, given by Jasper (1972) in his excellent review and compilation of surface tension data, show that, at low to moderate temperatures, the surface tension is essentially linear in temperature. In fact, Jasper (1972) provided constants for linear temperature dependence for most of the compounds included in his compilation and showed that the average deviation from the raw data was generally less than ±0.1 dyn/cm. The curves shown in Figure 8.1 from Eqs. 8.1 and 8.2 were obtained by fitting the data to obtain the constants σ_0 (the value of b was fixed at 1.222) or P^*, the parachor. Liquid density data were obtained from Kudchadker et al. (1978) for tetralin, Viswanath (1979) for quinoline, and TRCHP (1983) for 1-methylnaphthalene. At these relatively low temperatures, the vapor density needed in Eq. 8.2 is negligible. The fit of the data shown in Figure 8.1 demonstrates that, with accurate values of σ_0,

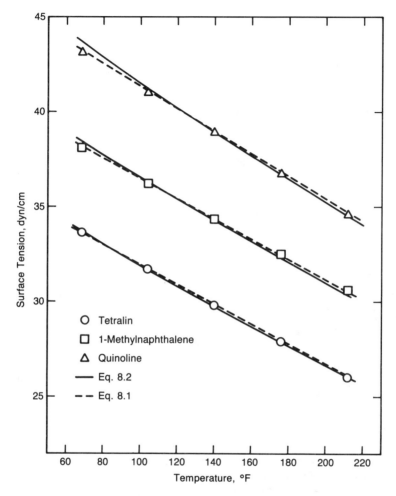

Figure 8.1   Low-temperature surface tension for selected model compounds.

parachor, and liquid and vapor density, both Eqs. 8.1 and 8.2 accurately represent the temperature dependence of the surface tension of the model compounds, at least over a moderate temperature range.

Auxiliary Correlations for Surface Tension Calculations

In Chapter 7, several liquid density correlations were shown to accurately predict the liquid density of coal-liquid-containing systems, with the Riedel correlation, Eq. 7.1, providing the most accurate predictions. Thus Eq. 7.1

can be used to predict liquid densities for surface tension calculations with Eq. 8.2. The vapor density can be predicted from the corresponding states correlation developed by Lee and Kesler (1975) for calculating PVT and thermodynamic properties of hydrocarbon and other nonpolar mixtures. The Lee-Kesler correlation, which was discussed briefly in Chapter 6 regarding the calculation of heat capacities, enthalpies, and heats of vaporization for liquid and vapor mixtures, requires that T_c, P_c, and ω be known (or estimated) for each component in a mixture in order to calculate a vapor density. For coal liquids, these "characterization" properties are predicted by the procedures recommended in Chapters 3 and 4. Thus with the critical properties and densities defined, one needs to estimate the parachor, P^* (or P^*/M), for Eq. 8.2, or to estimate σ_0 for Eq. 8.1, in order to calculate the surface tension of coal-liquid fractions.

CORRELATION OF PARAMETERS FOR SURFACE TENSION CALCULATIONS

Since the "molecular" structure of petroleum- or coal-derived liquids is generally not known, a group-contribution procedure such as that of Quayle (1953) for defined compounds cannot be used to predict parachor values for coal-liquid fractions. Thus more empirical correlations, based on normal boiling point and specific gravity, are needed for such predictions.

One such correlation is that developed by Sanborn and Evans and made available by Bondi (1971):

$$\sigma = \frac{673.7}{K_w}(1 - T_r)^{1.232} \qquad (8.3)$$

where K_w is the Watson characterization factor. Equation 8.3 shows that σ_0 is given by a simple function of K_w. Furthermore, if Eq. 8.3 is applied at some reference temperature, such as 60°F or the normal boiling point, the resulting surface tension value can be used to calculate a parachor value by rearranging Eq. 8.2 to

$$\frac{P^*}{M} = \frac{\sigma^{0.25}}{\rho^L - \rho^V} \qquad (8.4)$$

Two advantages of choosing 60°F instead of the normal boiling point are that the specific gravity at 60/60°F immediately provides a value of ρ^L and, for most coal liquid fractions of interest, ρ^V is negligible at 60°F, thus simplifying the calculation. This parachor value can be used at all temperatures because P^* is generally independent of temperature for nonpolar systems.

Another approach is to predict the parachor directly from the normal boiling point and specific gravity. Following an approach similar to that presented by Nokay (1959), the following correlation was developed for predicting P^*/M for a wide variety of hydrocarbons, including coal liquids:

$$P^*/M = 1.6652(T_b, °R)^{0.05873}(S)^{-0.64927} \tag{8.5}$$

Equation 8.5 predicts P^*/M for hydrocarbons ranging from n-paraffins to polynuclear aromatics with average errors of less than ±1%.

Figure 8.2 compares predicted surface tensions with experimental data

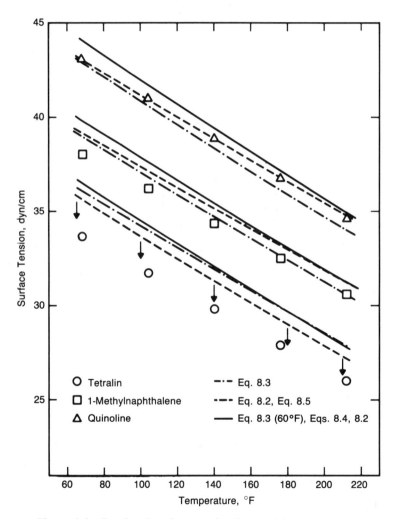

Figure 8.2   Predicted surface tension for model compounds.

for the three model compounds—tetralin, 1-methylnaphthalene, and quino-line—previously shown in Figure 8.1. These predictions were obtained by three different calculation schemes. In the first, Eq. 8.3 was used to calculate σ as a function of temperature; in the second, Eqs. 8.3 and 8.4 were used to calculate a value of (P^*/M), which was then used in Eq. 8.2. Finally, Eq. 8.5 was used to calculate (P^*/M) for Eq. 8.2. In all cases, the model compounds were characterized only by normal boiling point and specific gravity, and the liquid density was predicted by Eq. 7.1. Figure 8.2 shows that the most accurate surface tension predictions are obtained from Eq. 8.2 when the parachor is obtained from Eq. 8.5. The average errors for the three compounds are about 2.5% when Eq. 8.5 is used versus 4.1% for the approach utilizing Eqs. 8.3 and 8.4. However, all three procedures yield average errors of less than ±5% for the model compounds investigated, which is satisfactory for most process design calculations.

SURFACE TENSION CALCULATIONS FOR COAL-LIQUID MIXTURES

Relatively little attention has been given to the surface tension of mixtures, particularly those liquid mixtures containing supercritical components such as hydrogen or methane. This is true for both experimental work and, consequently, correlation development. Nevertheless, for practically all process design applications where surface tension calculations are of inter-est, these calculations must be performed for liquid mixtures containing supercritical components.

One of the most frequently used correlations for the surface tension of mixtures is that proposed by Weinaug and Katz (1943), which relates the mixture surface tension to the component parachors and mixture properties by

$$\sigma_m = \sum_i \left[P_i^* \left(\frac{x_i \rho^L}{M^L} - \frac{y_i \rho^V}{M^V} \right) \right]^4 \tag{8.6}$$

where ρ^L and ρ^V are the mass densities of the liquid and vapor phases, respectively, M^L and M^V are the corresponding molecular weights, and P_i^* is the parachor for component i. For coal-liquid fractions, P_i^* is obtained from Eq. 8.5.

The use of Eq. 8.6 for calculating the surface tension of mixtures generally assumes that the liquid and vapor phases of interest are at equilibrium with each other. Thus a VLE flash calculation must first be performed on the mixture in order to determine the appropriate phase compositions. These are used for both liquid and vapor density and mole-

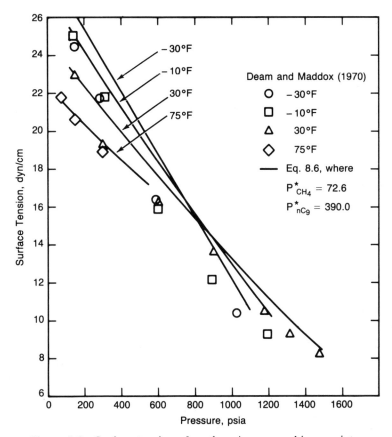

Figure 8.3   Surface tension of methane/n-nonane binary mixture.

cular weight calculations needed in Eq. 8.6. For the data analyses discussed in this chapter, all such VLE calculations were performed according to the recommendations given in Chapter 5.

The applicability of Eq. 8.6 to mixtures containing supercritical components is illustrated in Figure 8.3 for the data reported by Deam and Maddox (1970) on the methane/n-nonane binary mixture. The flash calculations were performed with the RKJZ equation of state, discussed in Chapter 5, with the binary parameter C_{ij} set equal to zero. The parachor values for methane and n-nonane were obtained from surface tension data for the pure components. Using the procedures described, Eq. 8.6 predicts these surface tension data with an average absolute error of about 7.9%. An alternative to Eq. 8.6 is presented in Chapter 10.

ANALYSIS OF DATA FOR COAL-LIQUID MIXTURES

The surface tension data reported by Hwang et al. (1982), Gray and Holder (1982), and Gray et al. (1983) include both wide-boiling range coal liquids (> 300°F) and "heart-cut" coal liquids (<20°F) which were pressurized with either hydrogen or methane. Thus, all of the mixtures evaluated contain supercritical components. However, none of these investigations determined the compositions of the liquid and vapor phases. Furthermore, they could not ascertain whether the systems were actually at equilibrium. In fact, in their analyses, Hwang et al. (1982) assumed that the liquid phase composition for all conditions was identical to the coal liquid feed and that the vapor phase was either pure hydrogen or pure methane. Clearly, this would not be true if equilibrium was attained, but might be a good approximation for the surface tension calculations if equilibrium was not achieved.

The data reported by Hwang et al. (1982) include measurements on most of the wide-boiling liquids for which liquid densities were also obtained and which were analyzed in Chapter 7. Thus the same procedure was used to characterize these liquids for the evaluation of their surface tension data. Each of the liquids was represented by the same 10 coal-liquid fractions used in the liquid density data analysis and given in Appendix B. The surface tension data reported by Gray and Holder (1982) and Gray et al. (1983) include measurements on the same narrow-boiling "heart-cut" fractions for which vapor-pressure data were obtained (and which were evaluated in Chapter 4). These "heart-cut" fractions, with typical boiling ranges of less than 20°F, were represented by single coal-liquid fractions. Thus the surface tension measurements reported for these "heart-cuts" can be considered as pseudobinary data, with supercritical hydrogen as the second component.

The analysis of the surface tension data reported by Hwang et al. (1982) with Eqs. 8.5 and 8.6 is summarized in Table 8.1. For the 68 data points with coal liquids in the presence of hydrogen, the overall average deviation in the predicted surface tension values is about 16.0%, which is about twice the stated uncertainty in the measurements. Essentially no difference was observed in the average deviations if the vapor phase was assumed to be hydrogen and the liquid phase was assumed to be the coal liquid. This was due to the very low hydrogen solubility in the liquid and the small quantity of coal liquid in the hydrogen vapor. Thus the two approaches to calculating the surface tension of the hydrogen-containing mixtures were essentially identical, and the possible lack of phase equilibrium at measurement conditions does not appear to be an issue for these systems.

This does not appear to be true for the surface tension data for the methane-containing systems. As shown in Table 8.1, large negative devia-

Table 8.1   Analysis of Surface Tension Data for EDS Coal Liquid
            Mixtures

EDS Liquids	No. Data Points	AAD[a]	Bias[a]
With Hydrogen			
IA-3	8	21.2	−10.2
IA-6	8	13.3	0.3
IA-10	8	20.1	7.5
IHS	20	12.3	10.0
WA-2	6	8.0	4.1
WA-5	6	22.4	−9.5
WA-6	6	19.5	−12.8
WV-1	6	17.0	−10.0
Total with H_2	68	16.0	−0.1
With Methane[b]			
WA-2	3	15.1 (4.2)	−15.1 (0.8)
WA-5	3	42.6 (13.4)	−42.6 (−13.4)
WA-6	3	47.2 (17.9)	−47.2 (−17.9)
WV-1	3	33.9 (11.7)	−33.9 (−7.1)
Total with CH_4	12	34.7 (11.8)	−34.7 (−9.4)
Total	80	18.8 (15.4)	−5.3 (−5.1)

[a] AAD: average absolute deviation, $[\sum |(\text{calc} - \text{exp})|/\text{exp}] \times 100/N$;
Bias: $[\sum (\text{calc} - \text{exp})/\text{exp}] \times 100/N$; N is number of data points.
[b] Numbers in parentheses obtained by assuming all methane is in vapor
phase, all coal liquids are in liquid phase.

tions occur for the predicted surface tensions of the methane-containing
systems when flash calculations are performed. This is due to the relatively
high solubility of methane in the coal-liquid mixtures, which reduces the
liquid density and liquid mixture parachor values. This results in a value of
the mixture surface tension that is much smaller than the value obtained by
assuming that all of the methane is in the vapor phase and all of the
coal-derived hydrocarbon is in the liquid phase. With this assumption of
nonequilibrium, the average deviation in the predicted surface tension is
reduced from 34.7% to 11.8%. This analysis suggests that, for mixtures
containing highly soluble supercritical components, the surface tension
values will depend strongly on the approach to phase equilibrium. Therefore, the corresponding phase compositions are needed in order to properly
interpret such measurements.

In addition to the uncertainty in the physical state of the two-phase mixtures, there appears to be a problem in the ability of Eq. 8.6 to predict the temperature and pressure dependence of the surface tension data. Figure 8.4 compares the experimental and predicted surface tension values for the EDS Coal Liquid IHS in the presence of hydrogen at several temperatures. The multiple points at each temperature correspond to different pressure values, with the highest calculated point corresponding to the lowest pressure (100 psia). Thus at constant temperature, the surface tension is overpredicted at low pressure (i.e., low hydrogen solubility), and the difference decreases with increasing pressure/solubility. Also note that, with increasing temperature, surface tension predictions are increasingly degraded and generally are the worst at the highest temperatures. This is also true for the other EDS liquids, except that most of the other liquids exhibit large negative deviations at the highest temperatures. Considering the apparent ability of Eq. 8.6 (and Eqs. 8.1 and 8.2) to predict the

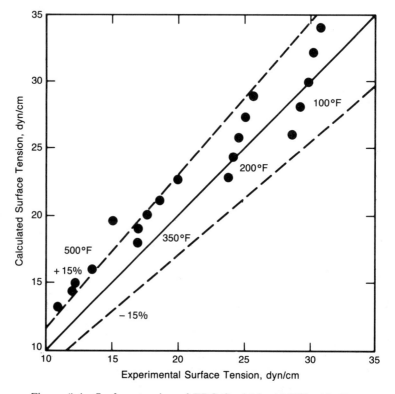

Figure 8.4   Surface tension of EDS Coal Liquid IHS with H_2.

temperature dependence of the surface tension of coal-liquid model compounds and of the methane/n-nonane binary mixture, these data appear to
be of questionable quality.

This is also true for the surface tension data reported by Gray et al.
(1983) and Gray and Holder (1982) on SRC-II "heart-cut" liquids in the
presence of hydrogen. The analysis of these data is summarized in Table
8.2 and shows even larger average deviations than for the EDS liquid
mixtures: about 20.1% with a bias of −8.7% versus 16.0% and −0.1%.
This result is especially disturbing since the SRC-II liquids are very narrow-
boiling (< 20°F) range materials and, in the presence of hydrogen, represent
systems which approach binary mixtures. This is not at all true for the very
wide-boiling (> 300°F) range EDS liquids, for which one might expect to
have difficulty in accurately predicting the mixture surface tension. Yet the
errors for the SRC-II liquid mixtures are larger than those for the EDS
liquids. Furthermore, assuming that hydrogen is present only in the vapor
phase and hydrocarbon is present only in the liquid phase does not
significantly improve the surface tension predictions.

Table 8.2 Analysis of Surface Tension Data for
SRC-II Coal Liquid Mixtures in Presence
of Hydrogen

SRC-II Liquids	No. Data Points	AAD[a]	Bias[a]
5HC[b]	3	34.1	−34.1
8HC	7	29.4	−27.4
11HC	5	22.8	−15.2
16HC	5	19.8	−5.6
17HC	5	16.6	3.7
4HC-B	2	12.9	−12.9
6HC[c]	4	21.4	−17.6
7HC-A	4	11.7	−10.3
10HC-A	4	13.5	−7.8
15HC-A	5	7.5	−2.1
18HC-A	5	26.2	26.2
Total	49	20.1	−8.7

[a]See Table 8.1.
[b]Gray et al. (1983)
[c]Gray and Holder (1982).

There also appears to be a serious problem regarding the temperature dependence of the SRC-II surface tension data. Figure 8.5 illustrates the relatively weak temperature dependence exhibited by the SRC-II liquid 8HC surface tension measurements versus the values calculated by Eq. 8.6. The predicted and experimental values are generally in good agreement at the lowest temperatures, but as temperature increases, Eq. 8.6 increasingly underpredicts the surface tensions, by as much as 50% at the highest temperatures. Gray and Holder (1982) stated that one should not expect the surface tensions of coal liquids to obey Eq. 8.6 (or Eqs. 8.1 and 8.2) because of their complex nature. However, the fact that model compounds (including heteroatom-containing compounds) and even "complex" petro-leum-derived fluids have been shown (American Petroleum Institute, 1983) to conform to these correlations raises questions about the data on coal liquids. Further experimental work is clearly warranted.

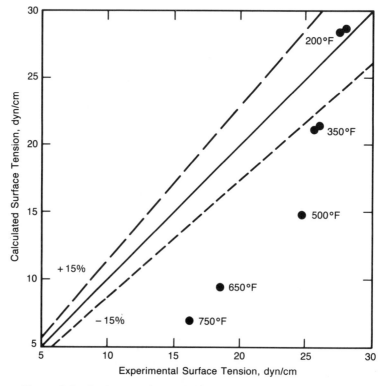

Figure 8.5   Surface tension of SRC-II Coal Liquid 8HC with H_2.

CONCLUSIONS AND RECOMMENDATIONS

The Weinaug-Katz correlation, Eq. 8.6, can be used for predicting the surface tension of mixtures containing coal-liquid fractions. The value of the parachor for each fraction can be obtained from either Eqs. 8.3 and 8.4 or Eq. 8.5. However, Eq. 8.5 appears to be slightly more accurate and has the additional advantage of not requiring a critical temperature.

Prior to the surface tension calculation with Eq. 8.6, a flash calculation must be performed to determine the equilibrium composition of the liquid and vapor phases. These flash calculations can be performed with the RKJZ equation of state, as recommended in Chapter 5. Once the liquid and vapor compositions are known, the liquid and vapor densities are calculated via the Riedel and Chueh-Prausnitz correlations, Eqs. 7.1 and 7.10, and the Lee-Kesler (1975) corresponding-states correlation, respectively.

For selected model compounds, these procedures result in predicted surface tensions with average deviations of about ±5% for temperatures below the compound's normal boiling point. Although very few data are available for these compounds at higher temperatures, larger errors in the predicted surface tensions should be expected. In addition, essentially no data exist for binary mixtures of model compounds with hydrogen or methane at any temperature. Such data would be very valuable for a more comprehensive analysis and development of surface tension correlations for coal-liquid mixtures.

Unfortunately, the results of the analysis of available surface tension data for mixtures of coal-liquid fractions with hydrogen or methane are inconclusive. Eq. 8.6 generally underpredicted the data reported by Hwang et al. (1982), Gray and Holder (1982), and Gray et al. (1983) by about 15 to 20%, with deviations frequently approaching −40 to −50% at the highest temperatures. Furthermore, the temperature dependence exhibited by these data is significantly different from that of a wide variety of hydrocarbon types, including heteroatom-containing model compounds and petroleum-derived fractions, for which the temperature dependence is adequately described by Eqs. 8.1, 8.2, and 8.6. More experimental work needs to be done to test the reliability of surface tension calculations, especially those for mixtures containing coal liquids and supercritical gases.

REFERENCES

American Petroleum Institute, *Technical Data Book—Petroleum Refining*, 4th Ed., Washington, DC, 1983, Chap. 10.

Bondi, A. (Shell Development), private communication, 1971.

Deam, J. R., and R. N. Maddox, Interfacial Tension in Hydrocarbon Systems, *J. Chem. Eng. Data*, **15**, 216 (1970).

Gray, J. A., and G. D. Holder, Selected Physical, Chemical, and Thermodynamic Properties of Narrow Boiling Coal Liquids from SRC-II Process, Supplementary Property Data, Report No. DOE/ET/10104-44, April 1982.

Gray, J. A., C. J. Brady, J. R. Cunningham, J. R. Freeman, and G. M. Wilson, Thermophysical Properties of Coal Liquids. 1. Selected Physical, Chemical, and Thermodynamic Properties of Narrow Boiling Range Coal Liquids, *Ind. Eng. Chem. Process Des. Dev.*, **22**, 410 (1983).

Hwang, S. C., C. Tsonopoulos, J. R. Cunningham, and G. M. Wilson, Density, Viscosity, and Surface Tension of Coal Liquids at High Temperatures and Pressures, *Ind. Eng. Chem. Process Des. Dev.*, **21**, 127 (1982).

Jasper, J. J., The Surface Tension of Pure Liquid Compounds, *J. Phys. Chem. Ref. Data*, **1**, 841 (1972).

Kudchadker, A. P., S. A. Kudchadker, and R. C. Wilhoit, *Tetralin*, API Publication 705, 1978.

Lee, B. I., and M. G. Kesler, A Generalized Thermodynamic Correlation Based on Three Parameter Corresponding States, *AIChE J.*, **21**, 510 (1975).

Macleod, D. B., On a Relation Between Surface Tension and Density, *Trans. Faraday Soc.*, **19**, 38 (1923).

Murad, S., Generalized Corresponding States Correlation for the Surface Tension of Liquids and Liquid Mixtures, *Chem. Eng. Commun.*, **24**, 353 (1983).

Nokay, R., Estimate Petrochemical Properties, *Chem. Eng.*, **66**(4), 147 (1959).

Quayle, O. R., The Parachors of Organic Compounds. An Interpretation and Catalogue, *Chem. Rev.*, **53**, 439 (1953).

Reid, R. C., J. M. Prausnitz, and T. K. Sherwood, *The Properties of Gases and Liquids*, 3rd Ed., McGraw-Hill, New York, 1977, Chap. 12.

Rice, P., and A. S. Teja, A Generalized Corresponding-States Method for the Prediction of Surface Tension of Pure Liquids and Liquid Mixtures, *J. Coll. Interface Sci.*, **86**, 158 (1982).

Sugden, S., The Variation of Surface Tension. VI. The Variation of Surface Tension with Temperature and Some Related Functions, *J. Chem. Soc.*, **125**, 32 (1924).

TRCHP (Thermodynamic Research Center Hydrocarbon Project), Selected Values of Properties of Hydrocarbons and Related Compounds, Texas A & M University, College Station, sheets extant 1983.

Viswanath, D. S., *Quinoline*, API Publication 711, 1979.

Weinaug, C. F., and D. L. Katz, Surface Tensions of Methane-Propane Mixtures, *Ind. Eng. Chem.*, **35**, 329 (1943).

Chapter 9

TRANSPORT PROPERTIES

Of the three transport properties—viscosity, thermal conductivity, and diffusivity—viscosity is the most important in process design calculations. It is used by itself (for example, in pressure drop calculations) but also in conjunction with thermal conductivity (in the Prandtl number) or diffusivity (in the Schmidt number). Our work focused on the viscosity of 1000°F− EDS coal liquids. Data for viscosity and thermal conductivity, which is the next most important transport property, have been reported for SRC-II coal liquids. In the case of diffusivity, apparently no published information exists on coal liquids.

The three transport properties are examined separately, but unevenly. There is much discussion on the viscosity of EDS and SRC-II coal liquids, some discussion on the thermal conductivity of SRC-II coal liquids, and little on diffusivity. In all cases, only the liquid transport properties are considered.

VISCOSITY

Most of the discussion on viscosity is based on the investigation of EDS coal liquids reported by Hwang et al. (1982). In addition, however, consideration is given to data on SRC-II coal liquids (Gray et al., 1983).

Correlations and Procedures Used

Two correlations were used by Hwang et al. in the analysis of the viscosity data. The first one, a "petroleum"-fraction liquid viscosity correlation developed by Abbott et al. (1971), is essentially an extension of a method proposed by Watson et al. (1935). Both Abbott and Watson used the K_w and the API gravity as parameters to obtain the kinematic viscosity at 100 and 210°F. However, Abbott's correlation was based on a large amount of experimental data for pure heavy components and crude fractions and is more accurate and suitable for computer use. The equation for kinematic viscosity, $\nu = \mu/\rho$ (where ρ is the mass density), is

$$\log_{10}\nu = a_0 + a_1 K_w + a_2(°API) + a_3(K_w)^2 + a_4(°API)^2 + a_5 K_w(°API)$$
$$+ \frac{b_1 K_w + b_2(°API) + b_3(K_w)^2 + b_4(°API)^2 + b_5 K_w(°API)}{c_0 + c_1 K_w + °API} \qquad (9.1)$$

The coefficients of Eq. 9.1 at 100 and 210°F are listed in Table 9.1. A graphical representation of the correlation is given in Figures 9.1 and 9.2.

The range of Abbott's correlation in the K_w/gravity plane is given in Figure 9.3. Abbott et al. noted that their correlation (1) should be used with care on the "edges" of the correlating parameters, particularly for $K_w < 10$ and $°API \le 0.0$, and (2) is most reliable for ν in the ranges 0.5 to 200 cSt (100°F) and 0.3 to 40 cSt (210°F).

A reexamination of Abbott's correlation (Hwang and Tsonopoulos, 1984) uncovered some problem areas that should be avoided because the correlation erroneously predicts that ν increases with increasing temperature. These problem areas are depicted in Figure 9.3 as regions A, B, and C. Region A is of special concern here because it includes polynuclear aromatic hydrocarbons and some coal liquid fractions.

Viscosities at other temperatures are obtained from a modification of the Walther (1930) equation:

$$\log \log(\nu + c) = a + b \log T \qquad (9.2)$$

where c is a function of viscosity; for $\nu > 1.5$ cSt, $c = 0.6$. Above 450°F, the left-hand side is assumed to be linear in T. The graphical version of the

Table 9.1 Coefficients of Equation 9.1

	100°F	210°F
a_0	4.39371	−0.463634
a_1	−1.94733	0.0
a_2	0.0	−0.166532
a_3	0.127690	0.0
a_4	3.26290×10^{-4}	5.13447×10^{-4}
a_5	-1.18246×10^{-2}	-8.48995×10^{-3}
b_1	0.0	8.03250×10^{-2}
b_2	10.9943	1.24899
b_3	0.171617	0.0
b_4	9.50663×10^{-2}	0.197680
b_5	−0.860218	0.0
c_0	50.3642	26.786
c_1	−4.78231	−2.6296

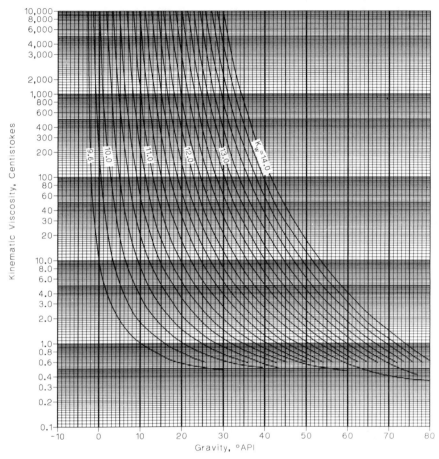

Figure 9.1   Abbott's correlation for the kinematic viscosity of fractions at 100°F (Eq. 9.1 and Table 9.1).

modified Walther equation is the ASTM standard viscosity chart. Figures 9.4 and 9.5 represent the low and high viscosity regions of the ASTM chart.

The correlation of Abbott et al. cannot directly predict the effect of pressure on viscosity. The correlation first calculates, independently of pressure, the kinematic viscosity for the fractions, and then introduces the effect of pressure only in the conversion of kinematic to dynamic viscosity, through the effect of pressure on the liquid density.

The effect of pressure on the dynamic viscosity can be estimated with the correlation of Kouzel (1965):

$$\log_{10} \frac{\mu}{\mu_0} = \frac{P - 14.696}{1000}(0.0239 + 0.01638\mu_0^{0.278}) \tag{9.3}$$

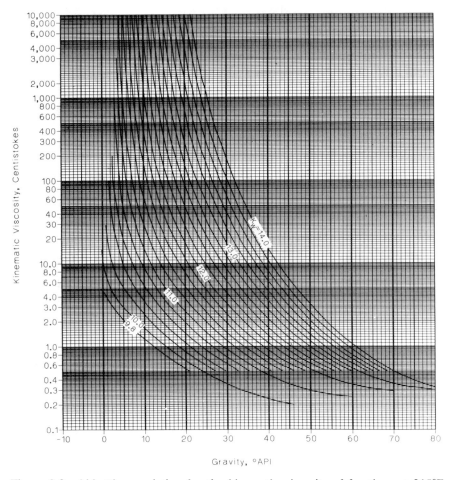

Figure 9.2 Abbott's correlation for the kinematic viscosity of fractions at 210°F (Eq. 9.1 and Table 9.1).

μ_0 is the viscosity of the liquid at the temperature of interest and atmospheric pressure or under its vapor pressure. Equation 9.3 is claimed to be reliable up to 5000 psi and 425°F.

Mixture viscosities for systems containing fractions can be calculated with a modification of Wright's (1946) method. This method requires viscosities at two temperatures for each component in the mixture. By plotting these data, or values at 100 and 210°F predicted with Eq. 9.1, on Figure 9.4 or 9.5, the slopes of the viscosity-temperature lines are obtained

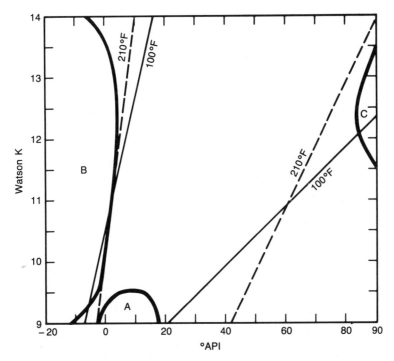

Figure 9.3 Range of Abbott's correlation in K_w/gravity plane. In regions A, B, and C the correlation erroneously predicts $\nu_{210} \geq \nu_{100}$.

and weight-averaged to obtain the slope of the viscosity-temperature curve of the mixture.

The second correlation used by Hwang et al. (1982) in the analysis of the viscosity data is an application of corresponding-states concepts to liquid viscosity. It was developed by Abbott and Kaufmann (1970) by analyzing a large amount of experimental data on defined compounds. This correlation is applicable from the freezing point to the critical point and has the general form

$$\ln \nu_r = \ln(\nu/\nu_c)$$

$$= A(\rho_r - 1) + B(\rho_r - 1)^{7/2} + \sum_{j=1}^{5} C_j[\exp(\rho - 1)^{(j+1)/2} - 1] \quad (9.4)$$

The reduced densities are calculated by Eq. 7.2. The coefficients in Eq. 9.4 are functions of acentric factor and hydrocarbon type; for example, paraffins, olefins, aromatics. For aromatics, which most closely represent the

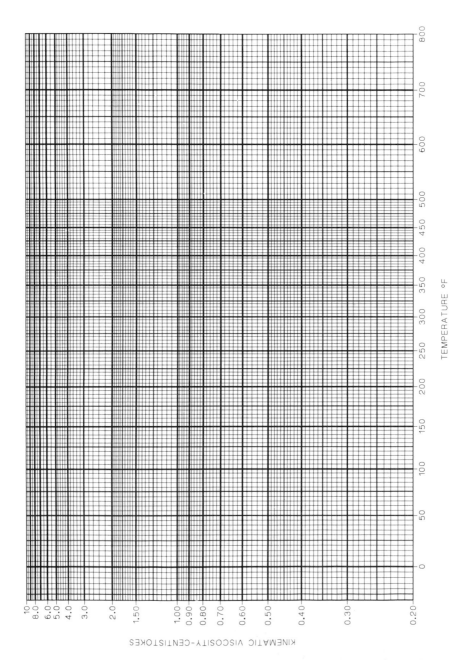

Figure 9.4 ASTM viscosity-temperature chart for low viscosity range. (Reprinted with permission from the American Society for Testing and Materials. Full-size charts are available from ASTM, 1916 Race Street, Philadelphia, PA 19103.)

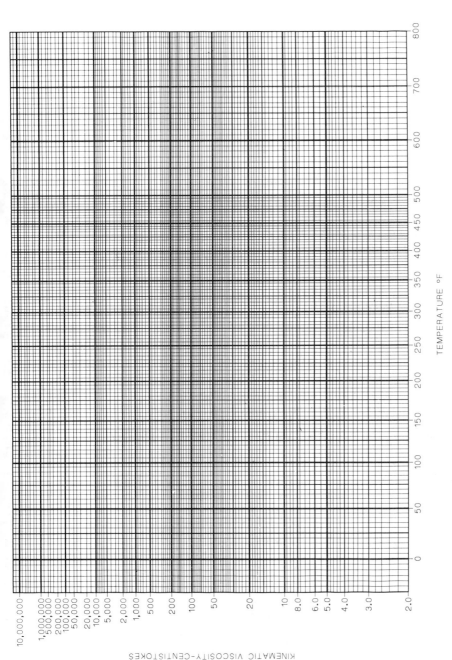

TEMPERATURE °F

KINEMATIC VISCOSITY-CENTISTOKES

Figure 9.5 ASTM viscosity-temperature chart for high viscosity range. (Reprinted with permission from the American Society for Testing and Materials. Full-size charts are available from ASTM, 1916 Race Street, Philadelphia, PA 19103.)

coal-liquid fractions, the coefficients are as follows:

$$A = 3.71104\omega^2 \tag{9.5a}$$

$$B = 0.25121 - 0.65882\omega \tag{9.5b}$$

$$C_1 = C_2 = 0 \tag{9.5c}$$

$$C_3 = 1.27195 \times 10^{-2} \tag{9.5d}$$

$$C_4 = -2.411 \times 10^{-4}\omega \tag{9.5e}$$

$$C_5 = 2.33529 \times 10^{-7}\omega^2 \tag{9.5f}$$

Critical kinematic viscositites required for converting the reduced kinematic viscosity are calculated from a straightforward extension of the correlation of Uyehara and Watson (1944):

$$\nu_c = 61.154 \times 10^{-4} \frac{T_c^{5/6}}{M^{1/2}} \left(\frac{RZ_c}{P_c}\right)^{1/3} \tag{9.6}$$

where ν_c is in cSt, T_c in K, and P_c in atm.

No single blending method for computing the viscosity of mixtures has been found applicable to all types of mixtures. The procedure used here for systems containing only defined compounds is the Kendall-Monroe (1917) method:

$$\mu_m = \left(\sum_i x_i \mu_i^{1/3}\right)^3 \tag{9.7}$$

If component i is supercritical, then its critical viscosity is used in Eq. 9.7.

Comparison with Experimental Data on EDS Coal Liquids

The nine EDS coal liquids used in our viscosity work (Hwang et al., 1982) are the same as those used in the investigation of liquid density (Chapter 7) and surface tension (Chapter 8). As discussed in Chapter 7, each of the coal-liquid cuts was represented by 10 narrow fractions. Inspection properties for the wide cuts are given in Appendix A, while the boiling point and specific gravity of the narrow fractions are given in Appendix B.

A comparison of calculated and experimental viscosities for coal liquid IHS at two isobars, 100 and 2000 psia, is shown in Figure 9.6. As discussed previously, the "petroleum"-fraction correlation cannot directly predict the effect of pressure on viscosity. The predicted small pressure effect on viscosity results from the pressure effect on density. Kouzel's correlation for the pressure effect on viscosity is considered separately.

As shown in Figure 9.6, the values predicted by the "petroleum"-fraction correlations are appreciably higher than experimental results at temperatures below 600°F. This disagreement is not unexpected, since the correlations (see Eq. 9.1 and Table 9.1) were based on petroleum fractions with °API's ranging from 10.6 to 54 and K_w's from 11.1 to 12.7 (Abbott et

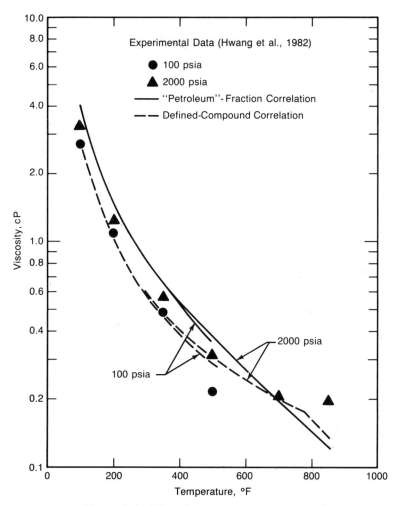

Figure 9.6 Viscosity of EDS Coal Liquid IHS.

al., 1971). The coal liquids of interest here have fractions with K_w's less than 11 and gravities less than 15 °API. Thus, a hazardous double extrapolation is involved in using this correlation, even when Region A in Figure 9.3 is avoided.

At 100 psia, the defined-compound correlation does very well for temperatures up to 350°F, but overpredicts the experimental data by 16% at 500°F. At 2000 psia, the defined-compound correlation underpredicts the viscosities, but does reasonably well up to 700°F. However, the blending procedure (Eq. 9.7) leads to a change in the slope of the curve at 790°F. This is because the system temperature is higher than the critical temperature of the most volatile component of the IHS coal liquid, and

Table 9.2 Viscosity of EDS Coal Liquids: Correlation Deviations[a]

| | "Petroleum"-Fraction Correlation | | | | Defined-Compound Correlation | | | |
| | Dev (cP) | | % Dev | | Dev (cP) | | % Dev | |
	AAD	Bias	AAD	Bias	AAD	Bias	AAD	Bias
Coal liquids without hydrogen (70 data points)	0.610	+0.601	45.4	+41.1	0.201	−0.172	14.1	−1.8
Coal liquids with hydrogen (20 data points)	0.715 (0.828)[b]	+0.696 (+0.827)	28.3 (35.4)	+26.9 (+35.1)	0.598 (0.422)	−0.598 (−0.419)	40.0 (23.4)	−40.0 (−20.8)
Total (90 data points)	0.633 (0.658)	+0.622 (+0.651)	41.6 (43.2)	+38.0 (+39.8)	0.289 (0.250)	−0.267 (−0.227)	19.9 (16.2)	−10.2 (−6.0)

[a] Dev = calc − exp; % Dev = $100 \times (\text{calc} - \text{exp})/\text{exp}$; AAD = $[\sum_{i=1}^{N} |(\text{Dev})_i|]/N$; Bias = $[\sum_{i=1}^{N} (\text{Dev})_i]/N$; N = number of data points.

[b] Values in parentheses are the results when hydrogen is ignored in the calculations.

166

therefore the critical viscosity of the component is used in computing the blend viscosity at all temperatures above 790°F.

A summary of the viscosity predictions for EDS coal liquids is given in Table 9.2. Although better results were obtained for the viscosity of hydrogen-free coal liquids when the coal-liquid fractions were simulated as defined compounds rather than as "petroleum" fractions, the reverse was true for hydrogen-containing coal liquids. This resulted from the blending procedure used in computing mixture viscosity. In the case of the defined-compound correlation, the critical viscosity of hydrogen was used in all the viscosity calculations of the coal liquids containing dissolved hydrogen. Apparently, this blending (Eq. 9.7) is poor for hydrogen/high-boiling compound mixtures.

To clarify the problem created by the blending procedure used in the viscosity predictions for hydrogen-containing coal liquids, calculations were also made by ignoring the hydrogen dissolved in the coal liquid. When this is done, the results of the calculations, also summarized in Table 9.2, indicate that the defined-compound correlation is markedly superior to the "petroleum"-fraction correlation. In the latter case, it is better to include hydrogen in the calculations.

Comparison of the experimental results with predictions indicates that both the "petroleum"-fraction and defined-compound correlations are not sufficiently accurate for viscosity predictions for coal-derived liquids. The average deviation ranges from 16% with defined-compound correlations, when hydrogen is ignored in the calculations, to 42% with "petroleum"-fraction correlations, when hydrogen is included in the calculations. Clearly, the defined-compound viscosity correlation, with the provision that hydrogen be ignored in the calculation, is the better model. The much larger error in the viscosity predictions with the "petroleum" correlation results from its use outside the original data base.

Viscosity Data on SRC-II Coal Liquids

Gray et al. (1983) presented liquid viscosity data up to 450°F on 16 SRC-II coal liquid cuts at approximately saturated conditions. They commented that the SRC-II results were in general agreement with the EDS data. Furthermore, Gray et al. examined a third correlation (Starling et al., 1980) that they found to be roughly equivalent to the Abbott-Kaufmann (1970) correlation; that is, the defined-compound correlation used in this chapter.

Our analysis of the SRC-II data is summarized in Table 9.3. Although the defined-compound correlation consistently underpredicts the viscosity of SRC-II coal liquids, overall it is better than the "petroleum"-fraction correlation. The reason is that the latter correlation breaks down for the heaviest cuts, 12 through 16 (Gray et al., 1983), which fall outside the $K_w/°API$ limits indicated in Figure 9.3. When these heavy cuts are

Table 9.3 Viscosity of SRC-II Coal Liquids: Correlation Deviations[a]

SRC-II Cut	No. of Data Points	"Petroleum"-Fraction Correlation				Defined-Compound Correlation			
		Dev(cP)		% Dev		Dev (cP)		% Dev	
		AAD	Bias	AAD	Bias	AAD	Bias	AAD	Bias
1	4	0.010	−0.008	5.3	−4.8	0.042	−0.042	19.3	−19.3
2	4	0.052	−0.052	17.6	−17.6	0.088	−0.088	25.1	−25.1
3	4	0.026	−0.010	9.1	−6.3	0.042	−0.042	12.7	−12.7
4	4	0.040	+0.008	9.3	−3.0	0.057	−0.057	14.0	−14.0
5	6	0.110	−0.110	9.6	−9.4	0.177	−0.177	22.0	−22.0
6	6	0.496	−0.405	33.2	−2.2	0.471	−0.471	29.3	−29.3
7	6	0.622	−0.493	34.9	+3.4	0.627	−0.627	28.9	−28.9
8	6	0.991	−0.905	26.5	−8.6	1.18	−1.18	40.3	−40.3
9	6	0.707	−0.451	30.4	+14.1	0.981	−0.981	28.3	−28.3
10	5	0.200	+0.119	25.9	+23.3	0.442	−0.442	25.3	−25.3
11	5	9.28	−9.28	46.5	−46.5	14.89	−14.89	66.2	−66.2
12[b]	5	11.8	+11.8	163.6	+163.6	6.11	−6.11	51.1	−51.1
13[b]	5	295.	+295.	1661.0	+1661.0	7.91	−7.91	49.4	−49.4
15[b]	5	4.84	+4.84	140.9	+140.9	2.64	−2.64	38.8	−38.8
16[b]	5	3476.	+3476.	9803.	+9803.	11.06	−11.06	54.6	−54.6
Total	76	250.0	+248.4	791.8	+770.8	3.116	−3.116	34.3	−34.3
Without cuts 12–16	56	1.169	−1.076	23.8	−4.6	1.753	−1.753	29.2	−29.2

[a]See Table 9.2.
[b]Outside the range of Eq. 9.1.

excluded, Table 9.3 shows that the "petroleum"-fraction correlation is the better method.

Effect of Pressure on Viscosity

Kouzel's correlation, Eq. 9.3, was used to estimate the pressure effect on the viscosity of EDS coal liquid IHS. The comparison in Figure 9.7 demonstrates that Kouzel's correlation is a good predictor up to 2000 psia.

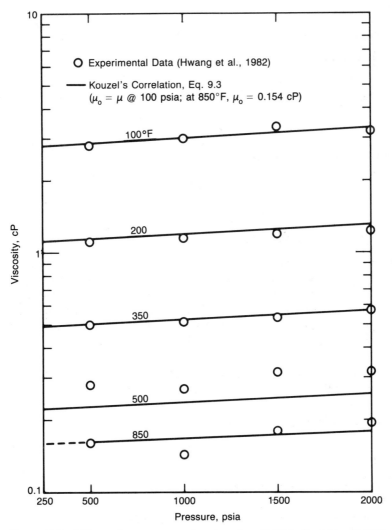

Figure 9.7   Effect of pressure on the viscosity of EDS Coal Liquid IHS.

The experimental results on EDS, SRC-II, and other coal liquids must be used to develop new correlations for the prediction of viscosity of coal-derived liquids. These new correlations should include the effect of pressure and must have better blending methods for gas/heavy liquid systems. For example, in place of Eq. 9.7, a mixing rule should be developed for the pseudocritical mixture kinematic viscosity, ν_{cm}, so that Eq. 9.4 can be applied to mixtures. In general, the prediction of the liquid viscosity of blends of heavy fractions is poor even when the viscosity of the fractions is known.

THERMAL CONDUCTIVITY

Much less is known about thermal conductivity, k_T, than the viscosity of coal liquids. Most of the available data are on SRC-II coal liquids (Gray et al., 1983), although some additional information has been presented by Baltatu et al. (1985) and others on SRC-I naphtha and COED distillate cuts.

In process design work, k_T is used in problems of heat conduction in flow systems through the Prandtl number

$$Pr = \frac{C_p \mu}{k_T}$$

Correlation of Thermal Conductivity

The thermal conductivity of petroleum fractions has been correlated with the gravity or the molecular weight of the fraction. Equally successful, however, has been a correlation that predicts the *same* thermal conductivity for all fractions at a given temperature, as long as the reduced temperature is below 0.85. Such a correlation is a testament to the low quality of both the data and the theories for liquid thermal conductivity.

Gray et al. (1983) examined the dependence of k_T on S, but found that k_T for some SRC-II liquids decreased with increasing S, while for others it increased. In view of this inconsistent behavior, the straightforward dependence of k_T on T (American Petroleum Institute, 1983) may be adequate for most engineering applications:

$$k_T = 0.07727 - 4.558 \times 10^{-5}(t) \tag{9.8}$$

where k_T is in Btu/[h · ft^2 · (°F/ft)] or Btu/(h · ft · °F). It should, of course, be remembered that Eq. 9.8 breaks down as the critical temperature is approached, where k_T goes to infinity.

An alternative to Eq. 9.8 is the corresponding-states method of Sato and

Riedel (Reid et al., 1977):

$$k_T = \frac{0.63862}{M^{0.5}} \left[\frac{3 + 20(1 - T_r)^{2/3}}{3 + 20(1 - T_{b,r})^{2/3}} \right] \tag{9.9}$$

Gray et al. (1983) regressed the experimental data on SRC-II coal liquids with a relationship between k_T and T_r recommended by Riedel (1955). When the T_c method recommended in Chapter 3 was used, the relationship between k_T and T_r is given by

$$k_T = 0.030918 + 0.058804(1 - T_r)^{2/3} \tag{9.10}$$

which is very similar to Gray's result.

Finally, for liquid mixtures, Reid et al. (1977) and the American Petroleum Institute (1983) recommend the method of Li (1976):

$$k_{T,m} = \sum_i \sum_j \Phi_i \Phi_j k_{T,ij} \tag{9.11a}$$

where

$$k_{T,ij} = 2(k_{T,i}^{-1} + k_{T,j}^{-1})^{-1} \tag{9.11b}$$

and the volumetric fraction Φ_i is defined as

$$\Phi_i = \frac{x_i v_i}{\sum_j x_j v_j} \tag{9.11c}$$

Comparison with Experimental Data

Table 9.4 summarizes the results of the analysis of the SRC-II data with Eqs. 9.8, 9.9, and 9.10.

Not surprisingly, Eq. 9.10, which is a fit of the data, gives the lowest deviation. What is surprising is that the direct relationship between k_T and temperature, Eq. 9.8, gives a lower deviation than the Sato-Riedel correlation, Eq. 9.9.

Although Eq. 9.10 is the best for the investigated SRC-II liquids, its range of applicability is undoubtedly very limited. For this reason, the simple Eq. 9.8 may still have some use as a rough predictive method. However, it should not be used for $T_r > 0.85$. Development of a more reliable and generalized correlation will require more data on coal liquid fractions and their mixtures (with H_2 and other gases) up to high reduced temperatures.

Table 9.4 Thermal Conductivity of Coal Liquids: Correlation Deviations[a]

SRC-II Cut	No. of Data Points	Equation 9.8			Equation 9.9			Equation 9.10		
		AAD	Bias	Max	AAD	Bias	Max	AAD	Bias	Max
2	6	21.2	+21.2	+28.3	12.1	+4.3	−20.0	3.8	+1.5	+6.3
4	6	14.1	+14.1	+18.9	7.8	+3.3	−11.1	2.8	+2.1	+6.1
6	12	4.7	−4.7	−7.6	7.2	−7.2	−13.8	6.4	−6.4	−10.3
8	12	2.6	−2.6	−6.4	7.0	−7.0	−13.7	1.6	−0.6	−2.8
10	12	2.3	−2.3	−6.0	8.4	−8.4	−12.9	2.5	+2.5	+4.9
12	12	6.0	−5.7	−12.1	14.1	−14.1	−18.4	2.7	+2.7	+5.0
16	6	8.7	−8.7	−13.0	17.6	−17.6	−19.6	1.4	+0.7	+4.2
18	8	12.0	−12.0	−14.4	23.0	−23.0	−25.3	2.2	+0.9	+5.9
Total	74	7.4	−1.6	+28.3	11.5	−9.2	−25.3	3.0	+0.2	−10.3

[a]See Table 9.2. Only % deviations are given.

DIFFUSIVITY

It is appropriate that diffusivity is the last property to be considered, as it generally is the least important property in process design. Not surprisingly, therefore, no literature information could be found on diffusion coefficients in coal liquids.

As noted in the API Technical Data Book (1983), diffusion coefficients are primarily used to correlate mass transfer properties through the Schmidt number (where ρ is the mass density)

$$Sc = \frac{\mu}{\rho D}$$

For isothermal mass transfer, the role of Sc is analogous to that of Pr in heat transfer. In the case of liquids, D is of the order of 10^{-4} ft^2/h (10^{-5} cm^2/s). The various methods recommended by Reid et al. (1977) or by the American Petroleum Institute (1983) are claimed to estimate D within 20%. Such estimates are entirely adequate for most engineering applications.

Two cases of diffusivity in liquids are of interest: (mostly) hydrocarbon liquids diffusing in (mostly) hydrocarbon liquids; and gases, such as H_2 and methane, diffusing into (mostly) hydrocarbon liquids.

For both cases, the API Technical Data Book (1983) recommends the following equation developed by Umesi (1980):

$$D_{1,2} = 5.922 \times 10^{-8} \left(\frac{T}{\mu_2}\right)\left(\frac{\bar{R}_2}{\bar{R}_1^{2/3}}\right) \tag{9.12}$$

$D_{1,2}$ is the diffusion coefficient for the infinitely dilute solute (component 1) in the solvent (component 2), in square feet per hour, μ_2 is the viscosity of the solvent, in centipoise, and $\bar{R}_i$ is the radius of gyration of component i, in angstrom units. $\bar{R}_i$ values for hydrocarbons are given in the API Technical Data Book. Values for a few gases are given below (Danner, 1984):

Gas	$\bar{R}_i$, Angstrom Units (10^{-10} m)
H_2	0.3708
N_2	0.5471
O_2	0.6800
CO_2	1.040

When the solute concentration is considerable, μ_2 in Eq. 9.12 should be replaced by the viscosity of the solution.

Experimental data for nonpolar liquids or gases diffusing into nonpolar liquids were reproduced to within 16% with Eq. 9.12. The only gases examined by Umesi, however, were methane, ethane, and propane.

There is much less information for inorganic gases diffusing in hydrocarbon liquids. Table 9.5 summarizes most of the available data, which were compiled by Akgerman and Gainer (1972), and predictions with the American Petroleum Institute (1983), Akgerman-Gainer, and Wilke-Chang (1955) methods. The comparison in Table 9.5 suggests that the Akgerman-Gainer method is the best for inorganic gases, but that method is complex and Umesi found it to be "very inaccurate" at high temperatures. Accordingly, the API method remains the choice, especially in view of its simplicity. Some data on the diffusion of H_2 and methane in polynuclear aromatic compounds are needed to better establish the predictive method of choice.

It is more important to obtain data at high temperature to establish the temperature variation of $D_{1,2}$. Equation 9.12 and most other correlations assume that

$$\frac{D_{1,2}\mu_2}{T} = \text{constant}$$

This may be a good assumption over a narrow temperature range, but not for extrapolating data at room temperature to $\sim 850°F$. The data of Peter and Weinert (1956) on the diffusion of H_2 in heavy paraffins at 106 and 220°C (223 and 392°F) suggest that $D\mu/T$ decreases with increasing temperature.

Another uncertainty in the prediction of $D_{1,2}$ is introduced by the

Table 9.5 Diffusion Coefficients of Inorganic Gases in Hydrocarbon Liquids

			$100 \times (\text{cal} - \text{exp})/\text{exp}$		
		$D_{1,2} \times 10^4$ (ft²/h) ($@$ t, °F)	API Method (Eq. 9.12)	Akgerman-Gainer	Wilke-Chang
Solute(1)	Solvent(2)				
H_2	n-Hexane	6.34 (77.72)	24.4	5	−18
H_2	Cyclohexane	2.74 (77.72)	−17.1	17	−36
N_2	Benzene	2.68 (77)	−11.4	−46	−41
O_2	Cyclohexane	2.06 (85.28)	−20.6	−27	−34
O_2	Benzene •	1.12 (85.28)	99.2	37	72
CO_2	n-Heptane	2.34 (77)	42.0	−30	7
	% average absolute deviation		35.8	27	35
	% bias		+19.4	−7	−8

uncertainty in μ. As a result, the uncertainty in $D_{1,2}$ at high temperatures may be closer to 50%. Even such uncertainty, however, may be acceptable for most applications in process design.

Finally, to apply Eq. 9.12 to coal liquids, the radius of gyration of the coal-liquid fraction(s) can be estimated with the following approximate relationship between $\bar{R}$ and ω:

$$\bar{R} = 1.0 + 9.5\omega \tag{9.13}$$

RECOMMENDATIONS

The defined-compound viscosity correlation, Eqs. 9.4–9.7, with the provision that dissolved light gases be ignored in the calculation, is recommended for predicting the viscosity of coal liquid fractions. The effect of pressure on the dynamic viscosity can be estimated with the correlation of Kouzel, Eq. 9.3.

The thermal conductivity of SRC-II coal liquids is reproduced very well with Eq. 9.10, but this equation may not have a wide applicability. For other liquids, the simple Eq. 9.8 may give a reasonable estimate. For estimating the thermal conductivity of liquid mixtures, the method of Li, Eq. 9.11, is recommended.

Finally, for diffusivity in coal liquid systems, the correlation developed by Umesi, Eq. 9.12, is recommended with the radius of gyration determined from Eq. 9.13. Very little is known, however, about diffusivity at high temperatures, and there are no published data on coal liquids.

REFERENCES

Abbott, M. M., and T. G. Kaufmann, Correlation of Orthobaric Kinematic Viscosities of Liquid n-Alkanes, *Can. J. Chem. Eng.*, **48**, 90 (1970).

Abbott, M. M., T. G. Kaufmann, and L. Domash, A Correlation for Predicting Liquid Viscosities of Petroleum Fractions, *Can. J. Chem. Eng.*, **49**, 379 (1971).

Akgerman, A., and J. L. Gainer, Predicting Gas-Liquid Diffusivities, *J. Chem. Eng. Data*, **17**, 372 (1972).

American Petroleum Institute, *Technical Data Book—Petroleum Refining*, 4th Ed., API, Washington, DC, 1983.

Baltatu, M. E., et al., Thermal Conductivity of Coal-Derived Liquids and Petroleum Fractions, *Ind. Eng. Chem. Process Des. Dev.* **24**, 325 (1985).

Danner, R. P. (Penn State U.), private communication (1984).

Gray, J. A., C. J. Brady, J. R. Cunningham, J. R. Freeman, and G. M. Wilson, Thermophysical Properties of Coal Liquids. 1. Selected Physical, Chemical, and Thermodynamic Properties of Narrow Boiling Range Coal Liquids, *Ind. Eng. Chem. Process Des. Dev.*, **22**, 410 (1983).

Hwang, S. C., and C. Tsonopoulos, A Correlation for Predicting Liquid Viscosities of Petroleum Fractions, *Can. J. Chem. Eng.*, **62**, 570 (1984).

Hwang, S. C., C. Tsonopoulos, J. R. Cunningham, and G. M. Wilson, Density, Viscosity, and Surface Tension of Coal Liquids at High Temperatures and Pressures, *Ind. Eng. Chem. Process Des. Dev.*, **21**, 127 (1982).

Kendall, J., and K. P. Monroe, The Viscosity of Liquids. II. The Viscosity-Composition Curve for Ideal Liquid Mixtures, *J. Am. Chem. Soc.*, **39**, 1787 (1917).

Kouzel, B., How Pressure Affects Liquid Viscosity, *Hydrocarbon Process.*, **44**(3), 120 (1965).

Li, C. C., Thermal Conductivity of Liquid Mixtures, *AIChE J.*, **22**, 927 (1976).

Peter, S., and M. Weinert, Diffusion Velocity of Hydrogen in Hydrocarbons at High Pressures, *Z. Phys. Chem.* (Frankfurt), **9**, 49 (1956), in German.

Reid, R. C., J. M. Prausnitz, and T. K. Sherwood, *The Properties of Gases and Liquids*, 3rd Ed., McGraw-Hill, New York, 1977.

Riedel, L., Compressibility, Surface Tension, and Thermal Conductivity in the Liquid Phase, *Chem. -Ing. -Tech.*, **27**, 209 (1955), in German.

Starling, K. E., et al., Coal-Calc Project Report to DOE, Coal Conversion Systems Technical Data Book Project, Institute of Gas Technology, Report OU/IGT/S-14366-1, August 1980.

Umesi, N. O., Correlating Diffusion Coefficients in Dilute Liquid Mixtures, M.S. thesis, The Pennsylvania State University, May 1980.

Uyehara, O. A., and K. M. Watson, A Universal Viscosity Correlation, *Nat. Petrol. News*, **36**, R-714 (1944).

Walther, C., The Viscosity-Temperature Diagram, *Pet. Z.*, **26**, 755 (1930), in German.

Watson, K. M., E. F. Nelson, and G. B. Murphy, Characterization of Petroleum Fractions, *Ind. Eng. Chem.*, **27**, 1460 (1935).

Wilke, C. R., and P. Chang, Correlation of Diffusion Coefficients in Dilute Solutions, *AIChE J.*, **1**, 264 (1955).

Wright, W. A., Prediction of Oil Viscosity on Blending, *ACS Div. Petrol. Chem. Preprints General Papers*, 71 (1946).

Chapter 10

FUTURE NEEDS
IN THERMODYNAMIC
AND TRANSPORT PROPERTIES

Chapters 3 through 9 have presented some of the methods currently used in the industry to characterize and predict the thermodynamic and transport properties of coal liquids. Some of these methods work well, others are poor. As has repeatedly been emphasized in this book, one major cause for the failures is that methods developed for petroleum liquids, which are mostly paraffinic and naphthenic, are used to predict the properties of coal liquids, which are mostly aromatic. In addition, coal liquids have a higher level of heteroatom-containing compounds than the petroleum liquids do. The very simple characterization of petroleum liquids, based on their boiling point and specific gravity, is much less satisfactory when the level of aromatics or heteroatoms is high.

A better characterization model is needed. Such a model should then be used as the basis for developing new property prediction methods. There are basic weaknesses in the current property prediction methods: those for defined compounds generally are corresponding-states methods based on perturbations around argon or methane, while those for (petroleum) fractions are limited to saturated hydrocarbons and low temperatures, generally less than ~400°F.

Before we do much more with theories, we need data on model compounds and coal liquids. The key problem is that we need data up to ~850°F, while most available experimental techniques and equipment can only take us up to ~400°F. New techniques and equipment are needed to make possible the measurement of properties up to ~850°F. At such temperatures, however, coal liquids, as well as the model compounds, become thermally unstable. It is necessary, therefore, that the measurements be carried out very quickly to minimize the effects of thermal degradation. Such methods are available today only for vapor pressure and VLE measurements, and are under development for thermal property measurements.

In the following sections, examples are given of the experimental and theoretical work that needs to be done to give us a better understanding of the thermodynamic and transport properties of coal liquids. The first section discusses the first step: characterization. Then follow sections on the thermodynamic and transport properties: what is needed and what is being done about it. In each section, an attempt has been made to summarize the results of recent or current work that appears to hold promise. It is neither an exhaustive nor a balanced treatment, but it is hoped that the following discussion suggests some promising directions for future work.

CHARACTERIZATION

The characterization information of interest here is only that needed for the prediction of the thermodynamic and transport properties of coal liquids. As in Chapter 3, we first consider the characterization of fractions and then that of defined compounds.

Fractions

The starting point in characterization and property measurements, and the correlation and prediction of these properties, should be narrow-boiling fractions rather than wide-boiling cuts. This will eliminate inconsistencies in the average boiling points recommended in Table 3.1. For example, the volume average is recommended for the correlation of heat capacity and the mean average for specific gravity. However, heat capacity is correlated with the specific gravity in the Watson-Nelson equation (Eq. 6.2). Narrow fractions, furthermore, often "behave" as single defined compounds, thus facilitating the transition from defined compounds to fractions.

For *distillates*, $t_b < 1050°F$, t_b and S are generally sufficient for *most* properties of hydrocarbon liquids.

The gas chromatographic distillation is the most convenient method for measuring the boiling point, especially since it gives the true boiling point, and thus no adjustments have to be made. This method is usable up to $t_b \cong 1000°F$, and some current work aims at raising that limit, as well as providing GC distillation standards for all types of liquids. Standards for coal liquids are needed for all analytical methods.

The boiling point provides a measure of the molecular size. An alternative would be the molecular weight, which can be measured cryoscopically or by vapor pressure osmometry. However, t_b and M should not be used together because they are highly intercorrelated. For distillates, t_b is definitely preferable to M because most separations take place by distillation, and knowledge of t_b is indispensable in distillation calculations.

A measure of molecular structure is provided by the specific gravity. This is measured directly at 60°F, which is convenient. However, 60°F can be significantly below the freezing point of defined compounds (see Table 1.1) or the pour point of fractions; in such cases, S would be an extrapolated value. It would be preferable to measure S a few degrees above the freezing point or at 60°F, whichever is higher. This is further discussed in the section on liquid density.

Alternatives to S are the fraction aromatic carbon (Alexander et al., 1985), the C/H ratio, or the %H. Gray (1984) found that these three alternative measures of structure were not superior to S in the correlation of bulk properties of heavy hydrocarbons. Alexander uses ^{1}H NMR to measure aromaticity. A better, but more expensive, method is ^{13}C NMR, which can also provide information on branchiness (fraction of carbon atoms which are in a terminal branch). Compound type analysis can readily be determined by mass spectrometry.

If the level of heteroatoms is high, the atomic formula becomes essential (especially to predict properties such as heat of combustion):

$$C_a H_b N_c O_d S_e$$

This is obtained by elemental analysis where, unfortunately, O is sometimes obtained by difference. In view of the importance of O in coal liquids, its level should be determined directly.

Elemental analysis may not suffice if the level of polar compounds is high. Then it may be important to distinguish between, for example, oxygen in -OH (phenols) and in -O- (ethers); see Alexander et al. (1985).

For *residua*, $t_b > 1050$°F, M must take the place of t_b because the latter generally is not measurable, although some current research has demonstrated that this upper limit for distillates can be raised ~200°F. Breaking down the residuum into fractions is possible by expressing the molecular weight of residua as a continuous distribution function (Gray, 1984).

Residua are fractionated in the laboratory by dissolving them in various solvents. For example, that portion of the residuum that is insoluble in *n*-heptane, or NHI, is commonly called asphaltenes. Asphaltenes are complex molecules that can have molecular weights in excess of 3000 (Speight and Moschopedis, 1979). Other investigators, however (e.g., McKay et al., 1978), claim that asphaltenes are smaller molecules with average $M \leq 1000$. This controversy underscores the current limitations in characterizing residua by solvent fractionation.

Creagh (1985) describes a method for characterizing residua that is used primarily in the petroleum industry. It involves sequential extraction of the residuum with *n*-heptane, toluene (which dissolves the aromatics), and pyridine (which dissolves practically all remaining organic material). In addition to all other tests also run on distillates (cryoscopic determination of the number-average molecular weight; elemental analysis; ^{1}H NMR spectroscopy to determine aromaticity; and IR spectroscopy to determine the concentration of -OH, -NH, -NH$_2$, and -CH$_3$ groups), Creagh discusses at length the use of gel-permeation chromatography to obtain an approximate molecular-weight distribution. Although GPC is a useful tool for characterizing residua, it provides only semiquantitative results because of limitations with polar compounds.

Defined Compounds

Critical constants are the key in most corresponding-states correlations for the properties of defined compounds. However, there may be a limit to their usefulness and to our ability to measure them at high temperatures. One question that has not been answered satisfactorily is whether property models for heavy liquids should depend on critical constants or on characterization parameters. If we are going to use critical constants for heavy liquids, we should determine whether critical constants approach limiting values as the carbon number goes to infinity. Similarly, if acentric factor is retained as the third parameter in corresponding-states models, we should establish its limiting value, if any, as the carbon number goes to infinity.

Bolotin et al. (1979), following an earlier observation by Kreglewski and Zwolinski (1961), presented generalized expressions for the properties of normal paraffins. The important feature of these expressions is that they predict the following limiting values for T_c and P_c for the infinitely large normal paraffin (carbon number $\rightarrow \infty$):

$$T_c = 1730°R$$

$$P_c = 0 \text{ psia}$$

It is also of interest that Bolotin's relationships lead to $P_c = 14.696$ psia, where $T_c = T_b$, for a carbon number of 76 ± 1. (The range is due to a slight inconsistency between the T_b, T_c, and P_c equations.)

The equation recommended in Chapter 3 for the prediction of critical temperature, Eq. 3.10, reduces to the line shown in Figure 10.1 for $t_c = t_b$. (For normal paraffins, represented by $S = 0.85$, the predicted critical temperature is 2133°R, which is higher than the Kreglewski-Zwolinski/Bolotin value of 1690°R.) However, at the same conditions, Eq. 3.15 predicts critical pressures much higher than 14.696 psia, as is also shown in Figure 10.1. Equation 3.14 reduces to

$$\log_{10} P_c = 2.22066 - 0.05445 K_w \tag{10.1}$$

at $t_c = t_b$. This equation gives $P_c = 14.696$ psia only for $K_w = 19.35$. In the K_w range of interest, Eq. 10.1 predicts much higher critical pressures (41.9 psia for $K_w = 11$; 53.8 psia for $K_w = 9$). What is needed to satisfy the

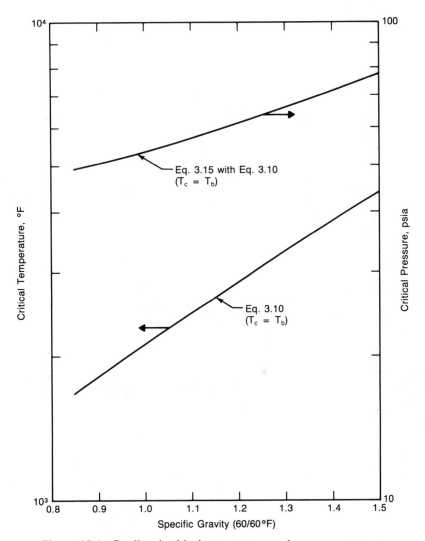

Figure 10.1   Predicted critical temperature and pressure at $t_c = t_b$.

condition $P_c = 14.696$ psia at $t_c = t_b$ is an equation of the form

$$\log_{10} P_c = 1.16720 + f'(T_b)\left[1 - \frac{T_b}{T_c}\right] + f''(S)\left[1 - \frac{T_b}{T_c}\right] \qquad (10.2)$$

Although Eq. 10.2 will give the right critical pressure at $t_b = t_c$, more extensive experimental data on t_c and P_c are needed for high-boiling compounds, such as those listed in Table 1.1, to make possible reliable extrapolations to even higher temperatures.

The measurement of critical constants, as well as of other properties, at high temperatures presents special difficulties because of the potential thermal decomposition. A method suitable for measuring the critical temperature of thermally unstable compounds has been described by Mogollon et al. (1982). A similar approach may also be feasible for the measurement of critical pressures.

Once t_c and P_c are known, the acentric factor, ω, must also be determined. This parameter has proven very useful in corresponding-states models, especially for $\omega < 0.5$. In the case of coal liquids, however, the heavy fractions may have ω's in excess of 1.0.* Data are needed on high-boiling compounds and fractions to determine how well the dependence of a property on ω, established with data for compounds with $\omega < 0.5$, extrapolates to $\omega > 1$. In anticipation of that, it is also important that the models used today have a reasonable limit for $\omega \gg 1$.

An example of what is meant by "reasonable limit" is provided by the relationship between Z_c, the critical compressibility factor, and ω. Many investigators assume that the following relationship is valid

$$Z_c = 0.2905 - 0.085\omega \qquad (10.3)$$

This relationship is specifically applicable to paraffins, but is also used for any nonpolar compound. Although it is reasonable for $\omega < 1$, Figure 10.2 clearly demonstrates that Eq. 10.3 has an unacceptable upper limit: Z_c decreases too rapidly with increasing ω and becomes negative for $\omega > 3.418$.

A better alternative to Eq. 10.3 is a relationship first presented in Chapter 7 (Eq. 7.9):

$$Z_c = \frac{1}{3.41 + 1.28\omega} \qquad (10.4)$$

This equation, which is also plotted on Figure 10.2, is similar to Eq. 10.3 up to $\omega = 0.8$, but predicts a much weaker dependence of Z_c on ω for $\omega > 1$ and does not go through 0.0.

Although Eq. 10.4 is an empirical relationship of little theoretical significance, it is clearly more "reasonable" than Eq. 10.3, which has a physically unacceptable upper limit. For example, Bolotin's (1979) generalized vapor pressure equation leads to $\omega = 4.66$ for a normal paraffin with 76

---

*The acentric factor, as developed by Pitzer, provides only a first-order correction for deviations from simple-fluid (e.g., Ar) behavior. As such, its theoretical significance for large molecules, where $\omega > 0.5$, is questionable; however, its importance and success as the third corresponding-states parameter even for large molecules have been firmly established.

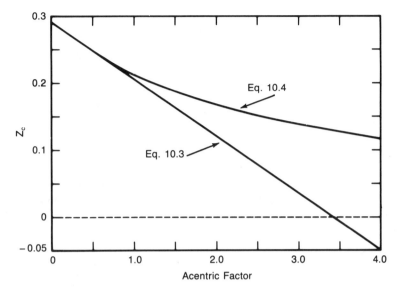

Figure 10.2 Predicted dependence of critical compressibility factor on acentric factor.

carbons. If this paraffin is assumed to have a critical density of $0.25 \, g/cm^3$, then $Z_c \cong 0.055$, while Eq. 10.4 predicts $Z_c = 0.107$, a much better choice than the negative value predicted by Eq. 10.3. Many other properties are assumed to have a linear dependence on ω (as in Eq. 10.3), but such a dependence is highly suspect for $\omega > 1$.

Finally, for polar fluids we need a fourth corresponding-states parameter. The dipole moment has proven useful as a fourth parameter for gases, but not for liquids. Gubbins and colleagues are currently working on a theory of polar liquids (Gray and Gubbins, 1984; Gubbins et al., 1984; Venkatasubramanian et al., 1984), and this research may lead to such a fourth parameter. One early result is that polarizability, α, must be taken into account along with the permanent dipole moment, μ, to give a better measure of the polarity of the liquid:

$$\mu' = \mu + \alpha \cdot f(\rho) \tag{10.5}$$

μ' is the effective dipole moment and $f(\rho)$ is a function of density that goes to zero as $\rho \to 0$.

In closing the discussion on the characterization of defined compounds, it should be emphasized that many investigators consider the classical law of corresponding states to be useless for large molecules. According to Prausnitz (1985), "the only corresponding-states theory that is sensible for large molecules is the Prigogine theory or one of its variations."

VAPOR PRESSURE

Chapter 4 has demonstrated that the Maxwell-Bonnell correlation is inadequate for coal-liquid fractions. Even the modification for $K_w < 12$ described in Chapter 4 is not highly accurate. Figures 10.3 and 10.4 illustrate the breakdown of the MB correlation for polar systems such as *m*-cresol and SRC-II Coal Liquid 6HC, which contains a high level of phenols. These figures strongly suggest that an additional parameter is needed for polar systems.

An alternative approach to the Maxwell-Bonnell or Lee-Kesler correlation was recommended by Smith et al. (1976). Their correlation, known as SWAP, requires one vapor pressure, just as MB does, and some "approximate characterization of molecular structure." For nonparaffinic hydrocarbons, SWAP uses parameters giving the aromaticity, naphthenicity, and branchiness (fraction of carbon atoms which are in a terminal branch) of the compound or fraction.

Gray (1984) made a limited comparison of SWAP against MB and found no significant improvement. Part of the problem with SWAP is the tem-

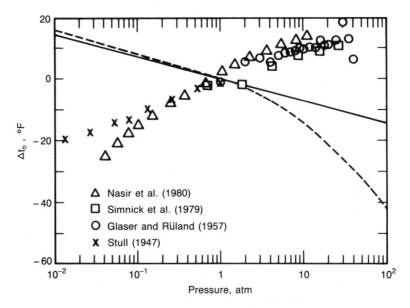

Figure 10.3 Boiling point correction for *m*-cresol with original (———) and modified (- - - - -) Maxwell-Bonnell correlation. (References for experimental data are given in Chapter 4.)

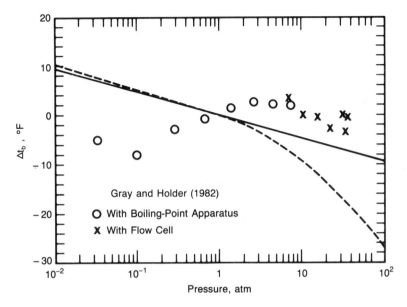

Figure 10.4   Boiling point correction for SRC-II Coal Liquid 6HC with original
(———) and modified (-----) Maxwell-Bonnell correlation.

perature dependence it uses for P^s:

$$\ln P^s = A + B/T + C/T^2 \qquad (10.6)$$

Equation 10.6 cannot accurately reproduce the temperature dependence
over a wide temperature range, and thus is unsuitable for extrapolating
above the normal boiling point or to very low pressures.

Macknick et al. (1978) modified SWAP by recommending a new relationship between t_b at 760 and t_b at 0.01 mm Hg. However, they applied their modified SWAP only to defined compounds. More recently,
Edwards et al. (1981) extended SWAP to include the effect of bound
nitrogen and sulfur. Then, Edwards and Prausnitz (1981a) used this
modification to extrapolate to higher temperatures their data on S-contain-
ing coal-derived compounds. The SWAP predictions of the normal boiling
point for compounds such as benzothiophene and dibenzothiophene showed
considerable error.

A new vapor pressure correlation is required. It should have a better
temperature dependence than Eq. 10.6 or the Antoine equation used by
MB. In addition, it should require a minimum of information in addition to
t_b and S. As noted earlier, the addition of the atomic formula may suffice.

Even with the atomic formula, however, it is unlikely that we could use group-contribution methods for fractions. An example of such a method is that of Macknick and Prausnitz (1979), who used the equation of Abrams et al. (1974):

$$\ln P^s = A + B/T + C \ln T + DT + ET^2 \tag{10.7}$$

This is much better than Eq. 10.6, especially for extrapolating to very low pressures. The coefficients of Eq. 10.7 were generalized by relating them to group contributions. [Edwards and Prausnitz (1981b) extended the group-contribution method of Macknick and Prausnitz (1979) to heavy defined hydrocarbons containing N or S.] Although the group contributions are readily available for defined compounds, it is not clear how they could be determined for incompletely characterized fractions. A method for the determination of functional groups in well-characterized coal liquids was proposed by Petrakis et al. (1983) and was applied to the SRC-II heavy distillate by Allen et al. (1984).

VAPOR–LIQUID EQUILIBRIA

The available high-temperature VLE data on model compounds with H_2 and methane may suffice, but the analysis in Chapter 5 suggests that more data are needed on mixtures of coal liquids with H_2 and methane. We also need VLE data on mixtures of model compounds or coal liquids with H_2O, NH_3, and H_2S. The liquid–liquid–vapor equilibrium behavior of H_2O with model compounds has been investigated by Heidman et al. (1985).

Cubic equations of state may be unreliable for residua. Better methods for heavy liquids may be the BACK equation of state (Chen and Kreglewski, 1977) or the perturbed-hard-chain model (Donohue and Prausnitz, 1978), which uses the Carnahan-Starling (1972) hard sphere equation of state to represent the repulsive forces. Vimalchand and Donohue (1985) developed the perturbed-anisotropic-chain model that is applicable to polar systems. Another method that uses the Carnahan-Starling model is the chain-of-rotators equation of state (Chien et al., 1983), which has also been presented in a cubic form (Lin et al., 1983; Kim et al., 1984).

Along with a better equation of state, better mixing rules than those described in Chapter 5 should improve predictions for polar systems. Most current work is on the so-called "density-dependent" mixing rules. These rules reduce to the well-known ones at low densities (gases), but take a different form at high densities (liquids). An example of the density-dependent mixing rules was presented by Luedecke and Prausnitz (1985; see also Mathias and Copeman, 1983; Prausnitz and Hu, 1984):

$$a_m = \sum_i \sum_j z_i z_j (a_i a_j)^{0.5}(1 - C_{ij}) + \frac{\rho}{RT} \sum_{i \neq j} \sum z_i z_j [z_i c_{i(j)} + z_j c_{j(i)}] \tag{10.8}$$

where $c_{i(j)}$ is a binary parameter that reflects noncentral forces when molecule j is infinitely dilute in molecules i.

Ongoing research on the BACK, perturbed-hard-chain (e.g., Wilhelm and Prausnitz, 1985), perturbed-anisotropic-chain, and chain-of-rotators equations of state, as well as on the density-dependent mixing rules, will most likely lead to VLE models that will be applicable to high-boiling and even strongly polar systems.

THERMAL PROPERTIES

As was noted in Chapter 6, predictions of liquid heat capacity with the modified Watson-Nelson correlation exhibit larger deviations as the concentration of oxygenated compounds, particularly phenolics, increases in the coal-derived liquid. The systematic deviation in liquid heat capacity with oxygen content is illustrated in Figure 10.5 for the series of SRC-II narrow-boiling fractions reported by Gray and Holder (1982). Figure 10.5 shows that the wt% oxygen in each of the fractions corresponds directly to the average % error in the liquid heat capacity as predicted by Eq. 6.4. Similar deviations were shown in Table 6.4 for coal liquid model compounds. Clearly, a third parameter is needed in addition to T_b and S to account for the effect of oxygen (or just the phenolic) concentration on liquid heat capacity.

Similarly, the association effects exhibited by high concentrations of phenolics in coal liquids result in poorly predicted heats of vaporization at low to moderate reduced temperatures. This is in part due to the poor prediction of the liquid heat capacity of these liquids in this range of reduced temperature. However, heat of vaporization predictions at $T_r = 0.8$ and above appear reasonably well behaved. Thus accounting for phenolic or oxygen content in the liquid heat capacity predictions will markedly improve the heat of vaporization predictions obtained with the procedures described in Chapter 6.

One final comment regarding both liquid heat capacity and heat of vaporization is the need for more high-temperature data for higher molecular weight and heteroatom-containing model compounds and fractions. Tables 6.1 and 6.2 illustrate that very few data are available for such compounds at temperatures above 400°F, primarily owing to their thermal instability. Therefore, experimental techniques are needed which would be capable of high-temperature (up to ~850°F) calorimetry, with very short residence times in order to minimize decomposition. One such potential technique is the "ballistic" method described by Mraw and Keweshan (1984), which is capable of measuring the total enthalpy change of a material from room temperature to temperatures as high as 930°F. Thus, a

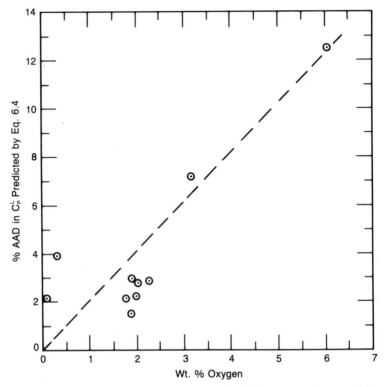

Figure 10.5 Effect of oxygen content on liquid heat capacity predictions for SRC-II fractions.

series of such measurements at increasing temperatures can potentially define the liquid heat capacity, heat of vaporization at $\sim t_b$, and the vapor heat capacity at superheated conditions for a compound or fraction of interest. More work is needed to extend the capabilities of the apparatus to pressures higher than ~ 1 atm.

Although reasonably reliable methods are available for predicting the heat of combustion of coal-liquid fractions, further improvements are needed to derive accurate values of the heat of formation for reaction enthalpy calculations. These correlations, which are based on the elemental analysis of the material, are generally more than adequate for predicting heating values of coal-liquid fuels. However, the calculation of reaction enthalpies typically involves relatively small differences between large enthalpy values of reactants and products. Therefore, the uncertainty in the heats of formation must be less than about 0.5% of the heat of combustion value in order to confidently predict reaction enthalpies. The high degree of accuracy will frequently require that the heat of combustion of selected materials be obtained experimentally.

Finally, although not discussed in Chapter 6, the Gibbs free energy of formation, ΔG_f, is needed for the prediction of simultaneous phase and chemical-reaction equilibria. This quantity is generally available for defined compounds, but no generalized correlations, such as those for ΔH_f, are available for predicting the free energy of formation of fractions. Furthermore, the group contribution procedures available for defined compounds (for both ΔH_f and ΔG_f) are unlikely to be applicable to fractions because the detailed structural information required by these correlations is generally not available. Thus alternate procedures are needed that may incorporate the elemental analysis and characterization parameters containing structural information such as aromaticity and branchiness.

LIQUID DENSITY

Chapter 7 demonstrated that, with an accurate reference density value and a reasonable estimate of the critical temperature, the liquid density can be predicted very accurately up to temperatures near the critical temperature for most of the compounds evaluated. Errors tend to increase for higher-boiling materials because of inaccuracies in the estimated critical temperatures and, perhaps more importantly, because the reference specific gravity at 60°F is far removed from the temperature range of interest. Furthermore, this specific gravity may be hypothetical if the material is below its freezing or pour point at 60°F. No unusual deviations were observed for the predicted liquid density of oxygen- or even phenolic-containing fractions or model compounds.

Improvements in the predicted liquid density of higher-boiling materials can be obtained by establishing a reference density/specific gravity at a temperature higher than 60°F and above the pour point. In addition, because of the strong temperature dependence of the liquid density in the vicinity of the critical point, it may be possible to establish an accurate value of the critical temperature from the liquid density model, whether it be the Riedel equation, Eq. 7.1, or the Rackett equation, Eq. 7.3. The heat of vaporization and surface tension, which also exhibit a strong temperature dependence as the critical temperature is approached, could be used with the liquid density model to establish an internally consistent value of the critical temperature. In general, more work could be done to derive characterization constants from a few selected property data and the appropriate physical property models.

Finally, although experimental techniques are available for measuring liquid densities at temperatures up to ~800°F, relatively few data have been reported for high-boiling materials. Such data are needed to improve both the molecular weight and temperature dependence of liquid density correlations. In addition, these data will make it possible to derive accurate characterization constants such as critical temperature and critical compressibility factor.

SURFACE TENSION

As noted in Chapter 8, there is a great need for data on the surface tension of coal liquid mixtures at high temperatures. Furthermore, such data should also include supercritical components such as hydrogen and methane, and should account for the distribution of the various components in the liquid and vapor phases. Data for both model compound systems and mixtures containing well-characterized, narrow-boiling fractions would be of great value in any subsequent correlation development effort.

The currently available data base is generally limited to surface tensions of defined compounds at low to moderate temperatures. These data typically exhibit a temperature dependence that is consistent with either Eq. 8.1 or Eq. 8.2, although the optimum exponent in Eq. 8.1 might range from 1.0 to 1.3. However, Gray et al. (1983) found that, in order to correlate their surface tension data for SRC-II liquids, a value of 0.4 was required for the exponent in Eq. 8.1. Such a mild temperature dependence is contrary to essentially all other available information and, therefore, requires further investigation.

Finally, as more reliable high-temperature surface tension data become available, consideration should be given to the thermodynamically rigorous treatment of the surface tension of mixtures. When the rigorous treatment is used, one result is (Sprow and Prausnitz, 1966)

$$\sum_{i=1}^{n} \left(\frac{x_i^B y_i^B}{\gamma_i^S}\right) \exp\left(\frac{\bar{A}_i \sigma_m - A_i \sigma_i}{RT}\right) = 1 \qquad (10.9)$$

where the superscript B denotes a bulk property and S denotes a surface property. $\bar{A}_i$ is the partial molar surface area of component i in the mixture, while A_i is the pure component property. Equation 10.9 was initially used by Sprow and Prausnitz in their investigation of nonpolar mixtures but was then extended to polar mixtures (Sprow and Prausnitz, 1967). However, Eq. 10.9 has not yet been applied to mixtures containing supercritical components or to undefined materials such as petroleum or coal liquids.

TRANSPORT PROPERTIES

Much more work is needed to improve or develop new experimental techniques for the measurement of transport properties of liquids at high temperatures. The National Bureau of Standards Center for Chemical Engineering is currently undertaking such an effort.

Perhaps the most general approach currently available for the calculation of the viscosity and thermal conductivity of nonpolar fluid mixtures is the extended corresponding-states model of Ely and Hanley (1981, 1983). This model is truly predictive in that it requires no transport property data as input. It is also very general in that it applies everywhere from the dilute gas to the dense liquid.

In the case of viscosity, for which the input parameters are the critical constants, the acentric factor, and the molecular weight of each pure component, Ely and Hanley (1981) examined a wide variety of C_1–C_{20} hydrocarbons and their mixtures. The claimed average deviation was 8% for pure components and 7% for mixtures.

Ely and Hanley (1983) used a similar approach in the prediction of the thermal conductivity of pure fluids and mixtures. One additional input property is required to predict the thermal conductivity: the ideal-gas heat capacity as a function of temperature, to account for the internal degrees of freedom for a polyatomic molecule. Experimental data for C_2–C_{24} hydrocarbons were predicted with an average deviation of about 7%, which is probably better than the average accuracy of the data.

Baltatu (1982) applied the Ely-Hanley model to petroleum fractions, for which only the boiling point and the specific gravity are assumed to be known. Baltatu examined liquid viscosity data for several crudes, but at temperatures only up to 212°F. The Ely-Hanley method was found to be within 7% of the data. Predictions for residua would probably be much less satisfactory. Indeed, several investigators have given up on predicting the viscosity of heavy fractions and instead recommend that it be measured at *three* temperatures so that the temperature dependence will also be established.

In view of its successful application on petroleum crudes, the Ely-Hanley method should be tested against coal liquids, especially at the high temperatures of interest in coal liquefaction. Such an application may test two possible limitations of the Ely-Hanley model: (1) polynuclear aromatic compounds are vastly different from methane, the reference fluid in the Ely-Hanley method; and (2) H_2/coal liquid mixtures will provide a severe test of the empirical correction used in the model to account for large differences in the critical volumes of the mixture components. In addition, the applicability of the Ely-Hanley method, which was developed for nonpolar fluids, to polar fluids can be tested by a comparison with data for phenols and some of the phenol-rich SRC-II fractions.

Finally, in the absence of data for coal liquids, and for virtually everything at high temperatures, there seems to be little need to work on the prediction of diffusion coefficients. However, any improvement in the prediction of viscosity may benefit the prediction of the temperature dependence of diffusion coefficients.

THERMODYNAMICS OF CONTINUOUS MIXTURES

The approach used in this book is based on dividing a coal liquid into several fractions or pseudocomponents according to their boiling points. An alternative approach is to treat the coal liquid as a continuous distribution function, so that its properties can be expressed in terms of the parameters of the distribution function selected to represent it.

Continuous distribution functions were examined by Edmister and Bowman in the early fifties (for example, Bowman and Edmister, 1951; Edmister and Bowman, 1952). The petroleum industry adopted some of the ideas of Edmister and Bowman (see also Redlich, 1960), but continued to rely on the use of discrete petroleum fractions. This may change as a result of the recent and significant resurgence of interest, especially in universities, in treating petroleum and coal liquids as continua rather than as mixtures of discrete components.

In the United States, two major groups working on "continuous thermodynamics" are those of E. D. Glandt at the University of Pennsylvania (e.g., Briano and Glandt, 1983) and J. M. Prausnitz at the Berkeley campus of the University of California (e.g., Cotterman and Prausnitz, 1985). Other investigators are J. M. Kincaid and G. Morrison of the National Bureau of Standards Center for Chemical Engineering (e.g., Gualtieri et al., 1982; Kincaid et al., 1983; Morrison and Kincaid, 1984), and S. A. Weil of the Institute of Gas Technology (Weil and Abbasian-Amin, 1984).

A major investigator outside of the United States is M. Rätzsch in East Germany (Rätzsch and Kehlen, 1983). Additional important contributions will undoubtedly come from other investigators working on the thermodynamics of continuous mixtures, both outside and within the United States.

"Continuous thermodynamics" may prove more successful in characterization and in describing bulk properties, such as the previously mentioned work on the molecular weight of residua (Gray, 1984), than in describing multicomponent VLE. In the latter case, it should be noted that Weil is relying entirely on continuous distribution functions in his VLE model, while Prausnitz is investigating the use of continuous distribution functions together with discrete components (Cotterman and Prausnitz, 1985).

THE NEED FOR DATA: A FINAL WORD

Experimental work will require at least five to ten years, especially because new techniques need to be developed for high-temperature work. As already noted, such methods are available for the measurement of vapor pressure and VLE, although even these methods need to be improved. Ballistic calorimetry (Mraw and Keweshan, 1984) has also been demonstrated to be a promising technique for measuring the heat capacity and heat of vaporization of organic liquids at high temperatures. Similar techniques are needed for measuring other properties at high temperatures.

We hope that the methods and the discussion in this book will help to initiate and carry out such an experimental effort. When the fruits of that work are collected and evaluated, it is certain that significantly better predictive tools than those in Chapters 3–9 will be developed. These tools, in turn, will be used to optimize the coal liquefaction processes of the future and to design coal liquefaction plants that will work, will be safe and of the right size, and will waste little energy.

REFERENCES

Abrams, D. S., H. A. Massaldi, and J. M. Prausnitz, Vapor Pressures of Liquids as a Function of Temperature. Two-Parameter Equation Based on Kinetic Theory of Fluids, *Ind. Eng. Chem. Fundam.*, **13**, 259 (1974); erratum, **16**, 392 (1977).

Alexander, G. L., A. L. Creagh, and J. M. Prausnitz, Phase Equilibria for High-Boiling Fossil-Fuel Distillates. 1. Characterization, *Ind. Eng. Chem. Fundam.*, **24**, 301 (1985).

Allen, D. T., L. Petrakis, D. W. Grandy, G. R. Gavalas, and B. C. Gates, Determination of Functional Groups of Coal-Derived Liquids by NMR and Elemental Analysis, *Fuel*, **63**, 803 (1984).

Baltatu, M. E., Prediction of the Liquid Viscosity for Petroleum Fractions, *Ind. Eng. Chem. Process Des. Dev.*, **21**, 192 (1982).

Bolotin, N. K., I. N. Zryakov, and A. M. Schelomentsev, Thermodynamic Properties of Heavy Hydrocarbons, *Russ. J. Phys. Chem.*, **53**, 812 (1979).

Bowman, J. R., and W. C. Edmister, Flash Distillation of an Indefinite Number of Components, *Ind. Eng. Chem.*, **43**, 2625 (1951).

Briano, J. G., and E. D. Glandt, Molecular Thermodynamics of Continuous Mixtures, *Fluid Phase Equilibria*, **14**, 91 (1983).

Carnahan, N. F., and K. E. Starling, Intermolecular Repulsions and the Equation of State for Fluids, *AIChE J.*, **18**, 1184 (1972).

Chen, S. S., and A. Kreglewski, Applications of the Augmented van der Waals Theory of Fluids. 1. Pure Fluids, *Ber. Bunsenges. Phys. Chem.*, **81**, 1048 (1977).

Chien, C. H., R. H. Greenkorn, and K. C. Chao, Chain-of-Rotators Equation of State, *AIChE J.*, **29**, 560 (1983).

Cotterman, R. L., and J. M. Prausnitz, Flash Calculations for Continuous or Semicontinuous Mixtures Using an Equation of State, *Ind. Eng. Chem. Process Des. Dev.*, **24**, 434 (1985).

Creagh, A. L., Characterization and Properties of Heavy Fossil Fuels, M.S. dissertation, University of California, Berkeley (1985).

Donohue, M. D., and J. M. Prausnitz, Perturbed-Hard-Chain Theory for Fluid Mixtures. Thermodynamic Properties of Mixtures in Natural Gas and Petroleum Technology, *AIChE J.*, **24**, 849 (1978).

Edmister, W. C., and J. R. Bowman, Equilibrium Conditions of Flash Vaporization of Petroleum Fractions, *Chem. Eng. Progr. Symp. Ser.*, **48**(3), 46 (1952).

Edwards, D. R., and J. M. Prausnitz, Vapor Pressures of Some Sulfur-Containing, Coal-Related Compounds, *J. Chem. Eng. Data*, **26**, 121 (1981a).

———, Estimation of Vapor Pressures of Heavy Liquid Hydrocarbons Containing Nitrogen or Sulfur by a Group-Contribution Method, *Ind. Eng. Chem. Fundam.*, **20**, 280 (1981b).

Edwards, D. R., C. G. Van de Rostyne, J. Winnick, and J. M. Prausnitz, Estimation of Vapor Pressures of High-Boiling Fractions in Liquefied Fossil Fuels Containing Heteroatoms Nitrogen or Sulfur, *Ind. Eng. Chem. Process Des. Dev.*, **20**, 138 (1981).

Ely, J. F., and H. J. M. Hanley, Prediction of Transport Properties. 1. Viscosity of Fluids and Mixtures, *Ind. Eng. Chem. Fundam.*, **20**, 323 (1981).

———, Prediction of Transport Properties. 2. Thermal Conductivity of Pure Fluids and Mixtures, *Ind. Eng. Chem. Fundam.*, **22**, 90 (1983).

Gray, C. G., and K. E. Gubbins, *Theory of Molecular Fluids*, Oxford University Press, 1984, Chap. 4.

Gray, J. A., C. J. Brady, J. R. Cunningham, J. R. Freeman, and G. M. Wilson, Thermophysical Properties of Coal Liquids. 1. Selected Physical, Chemical, and Thermodynamic Properties of Narrow Boiling Range Coal Liquids, *Ind. Eng. Chem. Process Des. Dev.*, **22**, 410 (1983).

Gray, J. A., and G. D. Holder, Selected Physical, Chemical and Thermodynamic Properties of Narrow Boiling Range Coal Liquids from the SRC-II Process, Supplementary Property Data, Report No. DOE/ET/10104-44, April 1982.

Gray, R. D., Jr., personal communication (1984).

Gualtieri, J. A., J. M. Kincaid, and G. Morrison, Phase Equilibria in Polydisperse Fluids, *J. Chem. Phys.*, **77**, 521 (1982).

Gubbins, K. E., Z. Maher, and M. C. Wojcik, New Developments in the Theory of Liquid Solutions, presented at the 1984 Annual AIChE Meeting, San Franscisco, November 25–30, 1984.

Heidman, J. L., C. Tsonopoulos, and G. M. Wilson, High-Temperature Mutual Solubilities of Hydrocarbons and Water. III. 1-Hexene, 1-Octene, and C_{10}–C_{12} Aromatics, Naphthenes, Paraffins, and Olefins, in preparation (1985).

Kim, H., K. C. Chao, and H. M. Lin, Cubic Chain-of-Rotators Equation of State for Supercritical and Polar Fluids, presented at the 1984 Annual AIChE Meeting, San Franscisco, November 25–30, 1984.

Kincaid, J. M., G. Morrison, and E. Lindeberg, Phase Equilibrium in Nearly Monodisperse Fluids, *Phys. Lett.*, **96A**, 471 (1983).

Kreglewski, A., and B. J. Zwolinski, A New Relation for Physical Properties of *n*-Alkanes and *n*-Alkyl Compounds, *J. Phys. Chem.*, **65**, 1050 (1961).

Lin, H. M., H. Kim, T. M. Guo, and K. C. Chao, Cubic Chain-of-Rotators Equation of State and VLE Calculations, *Fluid Phase Equilibria*, **13**, 143 (1983).

Luedecke, D., and J. M. Prausnitz, Phase Equilibria for Strongly Nonideal Mixtures from an Equation of State with Density-Dependent Mixing Rules, *Fluid Phase Equilibria*, **22**, 1 (1985).

Macknick, A. B., J. Winnick, and J. M. Prausnitz, Vapor Pressures of High-Boiling Liquid Hydrocarbons, *AIChE J.*, **24**, 731 (1978).

Macknick, A. B., and J. M. Prausnitz, Vapor Pressures of High-Molecular-Weight Hydrocarbons, *J. Chem. Eng. Data*, **24**, 175 (1979).

Mathias, P. M., and T. W. Copeman, Extension of the Peng-Robinson Equation of State to Complex Mixtures: Evaluation of the Various Forms of the Local Composition Concept, *Fluid Phase Equilibria*, **13**, 91 (1983).

McKay, J. F., et al., Petroleum Asphaltenes: Chemistry and Composition, *Adv. Chem. Ser.*, No. 170, 128 (1978).

Mogollon, E., W. B. Kay, and A. S. Teja, Modified Sealed-Tube Method for the Determination of Critical Temperature, *Ind. Eng. Chem. Fundam.*, **21**, 173 (1982).

Morrison, G., and J. M. Kincaid, Critical Point Measurements on Nearly Polydisperse Fluids, *AIChE J.*, **30**, 257 (1984).

Mraw, S. C., and C. F. Keweshan, Calvet-Type Calorimeter for the Study of High Temperature Processes. II. New Ballistic Method for the Enthalpy of Vaporization of Organic Materials at High Temperatures, *J. Chem. Thermodyn.*, **16**, 873 (1984).

Petrakis, L., D. T. Allen, G. R. Gavalas, and B. C. Gates, Analysis of Synthetic Fuels for Functional Group Determination, *Anal. Chem.*, **55**, 1557 (1983).

Prausnitz, J. M., private communication (1985).

Prausnitz, J. M., and Y. Hu, Equations of State from Generalized van der Waals Theory, presented at the 1984 Annual AIChE Meeting, San Francisco, November 25–30, 1984.

Rätzsch, M. T., and H. Kehlen, Continuous Thermodynamics of Complex Mixtures, *Fluid Phase Equilibria*, **14**, 225 (1983).

Redlich, O., A Distribution Function for Oil Mixtures and Polymers, *AIChE J.*, **6**, 173 (1960).

Smith, G., J. Winnick, D. S. Abrams, and J. M. Prausnitz, Vapor Pressures of High-Boiling Complex Hydrocarbons, *Can. J. Chem. Eng.*, **54**, 337 (1976).

Speight, J. G., and S. E. Moschopedis, Some Observations on the Molecular "Nature" of Petroleum Asphaltenes, *ACS Div. Petrol. Chem. Preprints*, **24**, 910 (1979).

Sprow, F. B., and J. M. Prausnitz, Surface Tension of Simple Liquid Mixtures, *Trans. Faraday Soc.*, **62**, 105 (1966).

———, Surface Thermodynamics of Liquid Mixtures, *Can. J. Chem. Eng.*, **45**, 25 (1967).

Venkatasubramanian, V., K. E. Gubbins, C. G. Gray, and C. G. Joslin, Induction Effects in Polar-Polarizable Liquid Mixtures: Calculation of Thermodynamic Properties Using Renormalized Perturbation Theory, *Mol. Phys.*, **52**, 1411 (1984).

Vimalchand, P., and M. D. Donohue, Thermodynamics of Quadrupolar Molecules: The Perturbed-Anisotropic-Chain Theory, *Ind. Eng. Chem. Fundam.*, **24**, 246 (1985).

Weil, S. A., and M. J. Abbasian-Amin, Fossil Fuel Process Oils as Continuous Fluids, presented at the AIChE Winter National Meeting, Atlanta, March 1984.

Wilhelm, A., and J. M. Prausnitz, Vapour Pressures and Saturated-Liquid Volumes for Heavy Fossil Fuel Fractions from a Perturbed Hard-Chain Equation of State, *Fuel*, **64**, 501 (1985).

INSPECTION DATA
ON EDS COAL LIQUIDS

All coal liquids described in this appendix were prepared at the Baytown laboratories of Exxon Research and Engineering Company. The inspection data were measured at the Baytown or Linden Laboratories of ER&E, at Brigham Young University, or at Wiltec Research Company.

Table A-1 presents the information on the six relatively narrow fractions used for vapor pressure measurements (Wilson et al., 1981); see Chapter 4. The five wider cuts used in the vapor-liquid equilibrium work (Wilson et al., 1981; Hwang et al., 1983) are described in Table A-2; see also Chapter 5. Finally, Table A-3 describes the seven cuts used, in addition to WA-5 and WA-6, in the physical property work (Hwang et al., 1982). WA-5, WA-6, and WV-1 were also used in the investigation of the heat capacity of coal liquids (Mraw et al., 1984).

With regard to the compound type, it should be noted that (a) all compounds that have at least one aromatic ring are counted as aromatics, and (b) a compound such as tetralin, for example, is counted as a *one*-ring aromatic compound.

The GC distillation data in Table A-2 exhibit much larger differences between laboratories than those in Table A-1 for the six narrow fractions. These differences are largely due to the wider boiling range of the cuts and the use of four laboratories (rather than only two for the narrow fractions). As indicated in Chapter 5, the GC distillation data selected for the vapor-liquid equilibrium calculations were those from Baytown, for I, II, and III, and Linden, for A-5 and A-6. Molecular weights are from BYU, except for III, for which the Baytown value was used.

In the case of the coal liquids in Table A-3, the molecular weight data selected were those from BYU. The only exception was IHS, for which the Linden value was used.

Table A-1 Inspection of Six EDS Coal Liquid Fractions[a]

Analysis	EDS Fraction					
	S-1[b]	S-2[b]	U-1[c]	U-2[c]	N-1[d]	N-2[d]
Elemental Analysis, wt%						
C	90.21	90.37	90.11	90.24	89.73	90.47
H	9.75	9.18	9.09	7.59	9.66	8.07
N	0.04	0.10	0.08	0.46	0.06	0.28
O	0.0	0.34	0.59	1.33	0.54[e]	1.14[f]
S	<0.001	0.01	0.13	0.38	0.0084	0.043
Compound Type Analysis, wt% by Mass Spectrometry						
Total saturates	18.3	18.2	12.3	6.5	22.97	23.66
Paraffins	0.5	3.6	2.8	2.3	7.92[g]	13.57[h]
1 Ring naphthenes	5.1	2.5	1.6	0.8	3.57	4.14
2 Ring naphthenes	9.1	5.2	3.8	1.4	4.89	2.59
3 Ring naphthenes	3.6	4.0	3.0	1.1	3.84	1.95
4+ Ring naphthenes	—	2.9	1.1	0.9	2.75	1.41
Aromatics	81.7	81.8	87.7	93.5	77.03	76.34
1 Ring	71.3	27.6	43.6	3.8	37.53[i]	3.90
2 Ring	10.4[j]	43.7	44.0	43.9	38.83	34.93
3 Ring	—	8.7	0.1	35.6	0.67	30.38
4+ Ring	—	1.8	—	10.2	0.0	7.13

GC Distillation, wt% Distilled @ °F	Baytown	BYU	BYU[k]	Baytown	BYU	Baytown[l]	Baytown	Baytown
1	356	357	465	407	417	556	397	573
5	369	366	493	434	438	585	421	591
10	380	377	506	444	446	596	429	600
20	393	390	519	450	450	610	436	612
50	406	405	549	473	476	655	458	655
80	415	420	588	497	500	694	485	694
90	424	424	609	510	510	714	501	713
95	433	425	627	521	522	731	510	730
99	457	448	646	551	543	780	530	763
Specific Gravity @ 60/60°F	—	—	—	—	—	—	0.9540	1.0360
Molecular Weight								
Baytown[m]	142		194	159		205	164	205
BYU[n]	134		179	158		225		

[a] Unless otherwise noted, all measurements were made at the Baytown laboratories of Exxon Research and Engineering Company; BYU = Brigham Young University.

[b] Narrow cut from Illinois No. 6 hydrogenated solvent.

[c] Narrow cut from Illinois No. 6 hydrogenated solvent.

[d] Narrow cut from Wyoming Wyodak unhydrogenated solvent.

[e] By difference; reported value was 0.45.

[f] By difference; reported value was 0.85.

[g] By difference; reported value was 7.67.

[h] By difference; reported value was 13.50.

[i] By difference; reported value was 35.53.

[j] By difference; reported value was 7.4.

[k] Baytown GC distillation discarded (t_b's 30 to 50°F higher than BYU results).

[l] BYU reported only that 56 wt% was distilled at about 646°F.

[m] Approximate M from mass spectrometric data.

[n] By freezing-point depression of benzene; accuracy is ±3%.

Table A-2 Inspections of Five EDS Coal Liquids Used in VLE Work[a]

Analysis	Coal Liquid I[b] Baytown	Coal Liquid II[c] Baytown	Coal Liquid II[c] Linden	Coal Liquid III[d] Baytown	Coal Liquid III[d] Linden	WA-5[e] Baytown	WA-5[e] Linden	WA-6[f] Baytown	WA-6[f] Linden
Elemental Analysis, wt%									
C	90.05	89.89	89.65	88.57	88.18	89.82	89.86[g]	90.46	89.22
H	9.57	9.64	9.65	8.19	8.25	9.50	9.03	9.14	9.25
N	0.07	0.11	<0.3	0.59	0.65	0.68	<0.3	0.40	0.37
O	0.29	0.34	—	2.15	—	0.0	—	0.0	—
S	0.015	0.018	—	0.50	—	0.0	—	0.0	—
Compound Type Analysis, wt% by Mass Spectrometry									
Total saturates	17.25	19.8		9.3		22.59		15.30	
Paraffins	2.17	2.2		1.0		10.12		3.03	
1 Ring naphthenes	3.64	5.0		2.1		3.72		2.89	
2 Ring naphthenes	4.88	5.7		2.7		3.68		3.91	
3 Ring naphthenes	3.83	4.0		2.0		2.86		3.07	
4+ Ring naphthenes	2.74	2.9		1.5		2.21		2.40	
Aromatics	82.75	80.2		90.7		77.41		84.70	
1 Ring	51.41	51.7		27.3		32.14		46.10	
2 Ring	25.43	23.2		35.1		34.16		27.35	
3 Ring	4.73	4.8		12.7		8.49		8.25	
4+ Ring	1.18	0.5		15.6		2.62		3.00	

	Baytown	BYU	Baytown	BYU	Baytown	BYU	Baytown	Linden	Wiltec	Baytown	Linden	Wiltec
GC Distillation, wt%												
Distilled @ °F												
1	387	387	356	356	367	361	357	350	340	344	350	350
5	401	396	378	375	386	386	370	380	375	356	380	378
10	407	400	389	395	404	405	383	390	390	369	390	400
20	422	406	410	406	433	430	402	411	405	392	407	408
50	504	460	483	475	580	577	457	478	445	456	488	472
80	613	588	595	575	830	—	569	607	547	580	605	590
90	674	646	668	631	912	—	629	696	600	662	658	650
95	728	703	732	675	972	—	684	752	630	740	690	690
99	831	805	800	725	1045	—	772	805	652	871	735	745
Specific Gravity @ 60/60°F	0.9715 (Baytown)		0.9645 (Linden)		1.055 (Linden)		0.9594 (Baytown) 0.9695 (Linden)			0.936 (Baytown) 0.9865 (Linden)		
Molecular Weight												
Baytown[h]	—		169		193		175			171		
Linden[i]	187		189		223		176			186		
BYU[j]	168		169		214		167			169		

[a] Unless otherwise noted, all measurements were made at the Baytown and Linden laboratories of Exxon Research and Engineering Company. BYU = Brigham Young University; Wiltec = Wiltec Research Company, Inc.

[b] Illinois No. 6 hydrogenated solvent.

[c] Illinois No. 6 hydrogenated solvent.

[d] Illinois No. 6 unhydrogenated "400–1000°F" cut.

[e] Wyoming Wyodak unhydrogenated solvent.

[f] Wyoming Wyodak hydrogenated solvent.

[g] Adjusted; reported value was 85.86.

[h] Approximate M from mass spectrometric data.

[i] By vapor-pressure osmometry.

[j] By freezing-point depression of benzene; estimated accuracy is ±3%.

Table A-3 Inspections of Seven EDS Coal Liquids Used in Physical Property Work[a]

				EDS Coal Liquid			
Analysis	IHS[b] Baytown	IA-10[c] Linden	IA-6[c] Linden	IA-3[c] Linden	IA-2[c] Linden	WV-1[a] Linden	WA-2[d] Linden
Elemental Analysis, wt%							
C	89.6	89.00	89.65	88.07	85.41	89.68	88.86
H	10.1	10.60	9.65	8.15	6.64	8.87	7.71
N	0.05	<0.3	<0.3	0.57	1.39	0.44	1.00
O	0.3	—	—	—	—	—	—
S	0.007	—	—	—	—	—	—
GC Distillation, wt%							
Distilled @ °F	Baytown	Linden	Linden	Linden	Linden	Linden	Linden
1	395	240	380	250	Not Available[e]	350	360
5	403	348	406	407		390	430
10	415	398	413	412		414	475
20	438	411	427	426		447	530
50	520	471	500	475		528	690
80	630	590	617	593		650	885
90	728	658	681	695		713	970
95	820	723	740	809		755	1020
99	955	790	790	975		805	1080

Specific Gravity @ 60/60°F (Linden)	0.9583	0.9359	0.9645	1.0067	1.2009	0.9975	1.0696
Molecular Weight							
Linden[f]	179	192	189	172	355	191	242
BYU[g]	—	164	172	167	482	192	242

[a]Measurements were made at the Baytown, for IHS, and Linden laboratories of Exxon Research and Engineering Company; M was also measured at Brigham Young University.

[b]Illinois No. 6 hydrogenated solvent; Baytown reported that IHS had 26.0 wt% saturates and 74.0 wt% aromatics, and $M = 174$.

[c]Various Illinois No. 6 cuts.

[d]Various Wyoming Wyodak cuts.

[e]Pour point higher than 250°F.

[f]By vapor-pressure osmometry (results for IA-10 and, especially, IA-2 are suspect).

[g]By freezing-point depression of benzene; estimated accuracy is ±3%.

REFERENCES

Hwang, S. C., C. Tsonopoulos, J. R. Cunningham, and G. M. Wilson, Density, Viscosity, and Surface Tension of Coal Liquids at High Temperatures and Pressures, *Ind. Eng. Chem. Process Des. Dev.*, **21**, 127 (1982).

Hwang, S. C., G. M. Wilson, and C. Tsonopoulos, Volatility of Wyoming Coal Liquids at High Temperatures and Pressures, *Ind. Eng. Chem. Process Des. Dev.*, **22**, 636 (1983).

Mraw, S. C., J. L. Heidman, S. C. Hwang, and C. Tsonopoulos, Heat Capacity of Coal Liquids, *Ind. Eng. Chem. Process Des. Dev.*, **23**, 577 (1984).

Wilson, G. M., R. H. Johnston, S. C. Hwang, and C. Tsonopoulos, Volatility of Coal Liquids at High Temperatures and Pressures, *Ind. Eng. Chem. Process Des. Dev.*, **20**, 94 (1981).

BOILING POINTS AND SPECIFIC
GRAVITIES OF FRACTIONS REPRESENTING
EDS COAL LIQUIDS

The wide EDS coal-liquid cuts used in the VLE and physical property work were represented by several narrow fractions, each one defined by its boiling point and specific gravity. This information is presented in Table B-1 for the fractions representing the "VLE cuts" and the "physical property cuts."

The information in Table B-1 suffices to estimate with the following equations all other properties needed in the calculations:

M: Eq. 3.5
t_c: Eq. 3.10
P_c: MB correlation at t_c; or Eq. 3.15 with Riedel equation
ω: Eq. 4.13 with MB correlation; or Eq. 4.15 with Riedel equation

Table B-1 Boiling Points and Specific Gravities (60/60°F) of Fractions Representing EDS Coal Liquids

Cumulative wt% Off	Coal Liquid I		Coal Liquid II		Coal Liquid III	
	t_b (°F)	S	t_b (°F)	S	t_b (°F)	S
0–5	394	0.8981	366	0.8819	376	0.9230
5–15	408	0.9075	388	0.8972	406	0.9442
15–25	422	0.9168	410	0.9120	437	0.9652
25–35	450	0.9346	432	0.9264	477	0.9911
35–45	477	0.9512	455	0.9410	525	1.020
45–55	505	0.9677	482	0.9575	582	1.052
55–65	535	0.9847	513	0.9757	650	1.088
65–75	570	1.004	552	0.9976	738	1.129
75–85	613	1.026	598	1.022	829	1.166
85–95	675	1.056	670	1.057	915	1.199
95–100	775	1.098	790	1.108	1015	1.233

Table B-1   (*Continued*)

Cumulative wt% Off	WA-5		WA-6		IHS	
	t_b (°F)	S	t_b (°F)	S	t_b (°F)	S
0–10	375	0.8978	375	0.9132	405	0.8885
10–20	402	0.9158	398	0.9289	425	0.9008
20–30	420	0.9274	419	0.9426	450	0.9156
30–40	444	0.9398	444	0.9584	478	0.9314
40–50	465	0.9547	472	0.9752	505	0.9460
50–60	490	0.9689	505	0.9940	535	0.9613
60–70	522	0.9863	540	1.013	567	0.9768
70–80	572	1.011	583	1.034	607	0.9952
80–90	650	1.046	629	1.055	668	1.020
90–100	750	1.085	690	1.081	825	1.075

Cumulative wt% Off	IA-10		IA-6		IA-3	
	t_b (°F)	S	t_b (°F)	S	t_b (°F)	S
0–10	325	0.8392	400	0.8999	356	0.9170
10–20	403	0.8920	420	0.9125	418	0.9601
20–30	416	0.9001	438	0.9236	433	0.9698
30–40	423	0.9044	460	0.9366	448	0.9792
40–50	457	0.9246	486	0.9513	465	0.9897
50–60	487	0.9414	516	0.9675	485	1.002
60–70	521	0.9594	550	0.9847	513	1.018
70–80	562	0.9797	591	1.004	556	1.040
80–90	621	1.006	646	1.028	636	1.079
90–100	725	1.047	735	1.062	825	1.151

Cumulative wt% Off	WV-1		WA-2	
	t_b (°F)	S	t_b (°F)	S
0–10	385	0.9113	425	0.9452
10–20	430	0.9405	505	0.9926
20–30	460	0.9587	552	1.017
30–40	485	0.9732	605	1.043
40–50	513	0.9888	660	1.067
50–60	545	1.005	720	1.091
60–70	580	1.023	780	1.113
70–80	625	1.043	850	1.135
80–90	678	1.066	925	1.157
90–100	755	1.095	1025	1.184

INDEX

A

Acentric factor:
  definition, 39, 54
  Edmister relation, 39
  mixing rules for, 134
  relation to critical compressibility
    factor, 130, 134, 182–183
  with Riedel vapor pressure equa-
    tion, 54–55
  theoretical significance, 182
Activity coefficient in surface ten-
    sion calculation, 190
Aging of coal liquids, 15
Ammonium salt deposition, 17
API gravity, 2
  conversion to specific gravity, 24
Aromaticity, measured by the Wat-
    son characterization fac-
    tor, 2–3
Aromatics:
  in coal liquids, 3
  in petroleum crude, 3
Asphaltenes, 179
ASTM viscosity–temperature
    charts, 162–163
Average boiling points of fractions,
    21–22
  Maxwell's suggestions for prop-
    erty correlation, 21–22
  mean average:
    definition, 22
    direct calculation, 22
  molar average, 22
  volume average, 21–22
  weight average, 21–22

B

Binary interaction parameters, *see*
    Mixing rules
Boiling point:
  average, *see* Average boiling
    points of fractions
  of model compounds, 8–9
    vs. GC retention time, 23
    vs. molecular weight, 32
    vs. specific gravity, 25
  of narrow coal-liquid fractions,
    60, 205–206
Boiling-point range of coal liquids,
    24, 80, 199, 201–202

C

Carbon-hydrogen ratio:
  in coal liquids, 3
  in petroleum crude, 3
Chao-Seader correlation, compari-
    son with RKJZ, 70–71, 75
Characterization:
  based on boiling point and spe-
    cific gravity, 2, 5, 113,
    178
  need for third parameter, 113,
    179, 187
  of coal liquids, 60, 80, 104–105
  of defined compounds, 180
  of distillates, 178
  measure of molecular size, 178
  measure of molecular structure,
    178–179